W9-BVR-844

SOURCE
The Prentice Hall
ENGINEERING SOURCE

Engineering Success

Peter Schiavone

University of Alberta

Prentice Hall
Upper Saddle River, NJ 07458

Library of Congress Information Available

Editor-in-chief: **MARCIA HORTON**
Acquisitions editor: **ERIC SVENDSEN**
Director of production and manufacturing: **DAVID W. RICCARDI**
Managing editor: **EILEEN CLARK**
Editorial/production supervision: **ROSE KERNAN**
Cover director: **JAYNE CONTE**
Creative director: **AMY ROSEN**
Marketing manager: **DANNY HOYT**
Manufacturing buyer: **PAT BROWN**
Editorial assistant: **GRIFFIN CABLE**

The author and publisher of this book have used their best efforts in
preparing this book. These efforts include the development, research,
and testing of the theories and programs to determine their effectiveness. The author and publisher shall not be liable in any event for inci-
dental or consequential damages in connection with, or arising out of,
the furnishing, performance, or use of these programs.
Printed in the United States of America

Printed with corrections July, 1999.

10 9 8 7 6 5 4 3 2

ISBN 0-13-080859-8

Prentice-Hall International (UK) Limited, *London*
Prentice-Hall of Australia Pty. Limited, *Sydney*
Prentice-Hall Canada, Inc., *Toronto*
Prentice-Hall Hispanoamericana, S.A., *Mexico*
Prentice-Hall of India Private Limited, *New Delhi*
Prentice-Hall of Japan, Inc., *Tokyo*
Prentice-Hall (Singapore) Pte., Ltd., *Singapore*
Editora Prentice-Hall do Brazil, Ltda., *Rio de Janeiro*

About ESource

The Challenge

Professors who teach the Introductory/First-Year Engineering course popular at most engineering schools have a unique challenge—teaching a course defined by a changing curriculum. The first-year engineering course is different from any other engineering course in that there is no real cannon that defines the course content. It is not like Engineering Mechanics or Circuit Theory where a consistent set of topics define the course. Instead, the introductory engineering course is most often defined by the creativity of professors and students, and the specific needs of a college or university each semester. Faculty involved in this course typically put extra effort into it, and it shows in the uniqueness of each course at each school.

Choosing a textbook can be a challenge for unique courses. Most freshmen require some sort of reference material to help them through their first semesters as a college student. But because faculty put such a strong mark on their course, they often have a difficult time finding the right mix of materials for their course and often have to go without a text, or with one that does not really fit. Conventional textbooks are far too static for the typical specialization of the first-year course. How do you find the perfect text for your course that will support your students educational needs, but give you the flexibility to maximize the potential of your course?

ESource—The Prentice Hall Engineering Source
http://emissary.prenhall.com/esource

Prentice Hall created ESource—The Prentice-Hall Engineering Source—to give professors the power to harness the full potential of their text and their freshman/first year engineering course. In today's technologically advanced world, why settle for a book that isn't perfect for your course? Why not have a book that has the exact blend of topics that you want to cover with your students?

More then just a collection of books, ESource is a unique publishing system revolving around the ESource website—http://emissary.prenhall.com/esource/. ESource enables you to put your stamp on your book just as you do your course. It lets you:

Control   You choose exactly what chapters or sections are in your book and in what order they appear. Of course, you can choose the entire book if you'd like and stay with the author's original order.

Optimize   Get the most from your book and your course. ESource lets you produce the optimal text for your student's needs.

Customize   You can add your own material anywhere in your text's presentation, and your final product will arrive at your bookstore as a professionally formatted text.

ESource Content

All the content in ESource was written by educators specifically for freshman/first-year students. Authors tried to strike a balanced level of presentation, one that was not either too formulaic and trivial, but not focusing heavily on advanced topics that most introductory students will not encounter until later classes. A developmental editor reviewed the books and made sure that every text was written at the appropriate level, and that the books featured a balanced presentation. Because many professors do not have extensive time to cover these topics in the classroom, authors prepared each text with the idea that many students would use it for self-instruction and independent study. Students should be able to use this content to learn the software tool or subject on their own.

While authors had the freedom to write texts in a style appropriate to their particular subject, all followed certain guidelines created to promote the consistency a text needs. Namely, every chapter opens with a clear set of objectives to lead students into the chapter. Each chapter also contains practice problems that tests a student's skill at performing the tasks they have just learned. Chapters close with extra practice questions and a list of key terms for reference. Authors tried to focus on motivating applications that demonstrate how engineers work in the real world, and included these applications throughout the text in various chapter openers, examples, and problem material. Specific Engineering and Science **Application Boxes** are also located throughout the texts, and focus on a specific application and demonstrating its solution.

Because students often have an adjustment from high school to college, each book contains several **Professional Success Boxes** specifically designed to provide advice on college study skills. Each author has worked to provide students with tips and techniques that help a student better understand the material, and avoid common pitfalls or problems first-year students often have. In addition, this series contains an entire book titled *Engineering Success* by Peter Schiavone of the University of Alberta intended to expose students quickly to what it takes to be an engineering student.

Creating Your Book

Using ESource is simple. You preview the content either on-line or through examination copies of the books you can request on-line, from your PH sales rep, or by calling(1-800-526-0485). Create an on-line outline of the content you want in the order you want using ESource's simple interface. Either type or cut and paste your own material and insert it into the text flow. You can preview the overall organization of the text you've created at anytime (please note, since this preview is immediate, it comes unformatted.), then press another button and receive an order number for your own custom book . If you are not ready to order, do nothing—ESource will save your work. You can come back at any time and change, re-arrange, or add more material to your creation. You are in control. Once you're finished and you have an ISBN, give it to your bookstore and your book will arrive on their shelves six weeks after the order. Your custom desk copies with their instructor supplements will arrive at your address at the same time.

To learn more about this new system for creating the perfect textbook, go to **http://emissary.prenhall.com/esource/**. You can either go through the on-line walkthrough of how to create a book, or experiment yourself.

Community

ESource has two other areas designed to promote the exchange of information among the introductory engineering community, the Faculty and the Student Centers. Created and maintained with the help of Dale Calkins, an Associate Professor at the University of Washington, these areas contain a wealth of useful information and tools. You can preview outlines created by other schools and can see how others organize their courses. Read a monthly article discussing important topics in the curriculum. You can post your own material and share it with others, as well as use what others have posted in your own documents. Communicate with our authors about their books and make suggestions for improvement. Comment about your course and ask for information from others professors. Create an on-line syllabus using our custom syllabus builder. Browse Prentice Hall's catalog and order titles from your sales rep. Tell us new features that we need to add to the site to make it more useful.

Supplements

Adopters of ESource receive an instructor's CD that includes solutions as well as professor and student code for all the books in the series. This CD also contains approximately **350 Powerpoint Transparencies** created by Jack Leifer—of University South Carolina—Aiken. Professors can either follow these transparencies as pre-prepared lectures or use them as the basis for their own custom presentations. In addition, look to the web site to find materials from other schools that you can download and use in your own course.

Titles in the ESource Series

Introduction to Unix
0-13-095135-8
David I. Schwartz

Introduction to Maple
0-13-095133-1
David I. Schwartz

Introduction to Word
0-13-254764-3
David C. Kuncicky

Introduction to Excel
0-13-254749-X
David C. Kuncicky

Introduction to MathCAD
0-13-937493-0
Ronald W. Larsen

Introduction to AutoCAD, R. 14
0-13-011001-9
Mark Dix and Paul Riley

Introduction to the Internet, 2/e
0-13-011037-X
Scott D. James

Design Concepts for Engineers
0-13-081369-9
Mark N. Horenstein

Engineering Design—A Day in the Life of Four Engineers
0-13-660242-8
Mark N. Horenstein

Engineering Ethics
0-13-784224-4
Charles B. Fleddermann

Engineering Success
0-13-080859-8
Peter Schiavone

Mathematics Review
0-13-011501-0
Peter Schiavone

Introduction to C
0-13-011854-0
Delores M. Etter

Introduction to C++
0-13-011855-9
Delores M. Etter

Introduction to MATLAB
0-13-013149-0
Delores M. Etter with David C. Kuncicky

Introduction to FORTRAN 90
0-13-013146-6
Larry Nyhoff and Sanford Leestma

About the Authors

No project could ever come to pass without a group of authors who have the vision and the courage to turn a stack of blank paper into a book. The authors in this series worked diligently to produce their books, provide the building blocks of the series.

Delores M. Etter is a Professor of Electrical and Computer Engineering at the University of Colorado. Dr. Etter was a faculty member at the University of New Mexico and also a Visiting Professor at Stanford University. Dr. Etter was responsible for the Freshman Engineering Program at the University of New Mexico and is active in the Integrated Teaching Laboratory at the University of Colorado. She was elected a Fellow of the Institute of Electrical and Electronic Engineers for her contributions to education and for her technical leadership in digital signal processing. IN addition to writing best-selling textbooks for engineering computing, Dr. Etter has also published research in the area of adaptive signal processing.

Sanford Leestma is a Professor of Mathematics and Computer Science at Calvin College, and received his Ph.D from New Mexico State University. He has been the long time co-author of successful textbooks on Fortran, Pascal, and data structures in Pascal. His current research interests are in the areas of algorithms and numerical computation.

Larry Nyhoff is a Professor of Mathematics and Computer Science at Calvin College. After doing bachelors work at Calvin, and Masters work at Michigan, he received a Ph.D. from Michigan State and also did graduate work in computer science at Western Michigan. Dr. Nyhoff has taught at Calvin for the past 34 years—mathematics at first and computer science for the past several years. He has co-authored several computer science textbooks since 1981 including titles on Fortran and C++, as well as a brand new title on Data Structures in C++.

Acknowledgments: We express our sincere appreciation to all who helped in the preparation of this module, especially our acquisitions editor Alan Apt, managing editor Laura Steele, development editor Sandra Chavez, and production editor Judy Winthrop. We also thank Larry Genalo for several examples and exercises and Erin Fulp for the Internet address application in Chapter 10. We appreciate the insightful review provided by Bart Childs. We thank our families—Shar, Jeff, Dawn, Rebecca, Megan, Sara, Greg, Julie, Joshua, Derek, Tom, Joan; Marge, Michelle, Sandy, Lori, Michael—for being patient and understanding. We thank God for allowing us to write this text.

Mark Dix began working with AutoCAD in 1985 as a programmer for CAD Support Associates, Inc. He helped design a system for creating estimates and bills of material directly from AutoCAD drawing databases for use in the automated conveyor industry. This system became the basis for systems still widely in use today. In 1986 he began collaborating with Paul Riley to create AutoCAD training materials, combining Riley's background in industrial design and training with Dix's background in writing, curriculum development, and programming. Dix and Riley have created tutorial and teaching methods for every AutoCAD release since Version 2.5. Mr. Dix has a Master of Arts in Teaching from Cornell University and a Masters of Education from the University of Massachusetts. He is currently the Director of Dearborn Academy High School in Arlington, Massachusetts.

Paul Riley is an author, instructor, and designer specializing in graphics and design for multimedia. He is a founding partner of CAD Support Associates, a contract service and professional training organization for computer-aided design. His 15 years of business experience and 20 years of teaching experience are supported by degrees in education and computer science. Paul has taught AutoCAD at the University of Massachusetts at Lowell and is presently teaching AutoCAD at Mt. Ida College in Newton, Massachusetts. He has developed a program, Computer-Aided Design for Professionals that is highly regarded by corporate clients and has been an ongoing success since 1982.

David I. Schwartz is a Lecturer at SUNY-Buffalo who teaches freshman and first-year engineering, and has a Ph.D from SUNY-Buffalo in Civil Engineering. Schwartz originally became interested in Civil engineering out of an interest in building grand structures, but has also pursued other academic interests including artificial intelligence and applied mathematics. He became interested in Unix and Maple through their application to his research, and eventually jumped at the chance to teach these subjects to students. He tries to teach his students to become incremental learners and encourages frequent practice to master a subject, and gain the maturity and confidence to tackle other subjects independently. In his spare time, Schwartz is an avid musician and plays drums in a variety of bands.

Acknowledgments: I would like to thank the entire School of Engineering and Applied Science at the State University of New York at Buffalo for the opportunity to teach not only my students, but myself as well; all my EAS140 students, without whom this book would not be possible—thanks for slugging through my lab packets; Andrea Au, Eric Svendsen, and Elizabeth Wood at Prentice Hall for advising and encouraging me as well as wading through my blizzard of e-mail; Linda and Tony for starting the whole thing in the first place; Rogil Camama, Linda Chattin, Stuart Chen, Jeffrey Chottiner, Roger Christian, Anthony Dalessio, Eugene DeMaitre, Dawn Halvorsen, Thomas Hill, Michael Lamanna, Nate "X" Patwardhan, Durvejai Sheobaran, "Able" Alan Somlo, Ben Stein, Craig Sutton, Barbara Umiker, and Chester "JC" Zeshonski for making this book a reality; Ewa Arrasjid, "Corky" Brunskill, Bob Meyer, and Dave Yearke at "the Department Formerly Known as ECS" for all their friendship, advice, and respect; Jeff, Tony, For-

rest, and Mike for the interviews; and, Michael Ryan and Warren Thomas for believing in me.

Ronald W. Larsen is an Associate Professor in Chemical Engineering at Montana State University, and received his Ph.D from the Pennsylvania State University. Larsen was initially attracted to engineering because he felt it was a serving profession, and because engineers are often called on to eliminate dull and routine tasks. He also enjoys the fact that engineering rewards creativity and presents constant challenges. Larsen feels that teaching large sections of students is one of the most challenging tasks he has ever encountered because it enhances the importance of effective communication. He has drawn on a two year experince teaching courses in Mongolia through an interpreter to improve his skills in the classroom. Larsen sees software as one of the changes that has the potential to radically alter the way engineers work, and his book Introduction to Mathcad was written to help young engineers prepare to be productive in an ever-changing workplace.

Acknowledgments: To my students at Montana State University who have endured the rough drafts and typos, and who still allow me to experiment with their classes— my sincere thanks.

Peter Schiavone is a professor and student advisor in the Department of Mechanical Engineering at the University of Alberta. He received his Ph.D. from the University of Strathclyde, U.K. in 1988. He has authored several books in the area of study skills and academic success as well as numerous papers in scientific research journals.

Before starting his career in academia, Dr. Schiavone worked in the private sector for Smith's Industries (Aerospace and Defence Systems Company) and Marconi Instruments in several different areas of engineering including aerospace, systems and software engineering. During that time he developed an interest in engineering research and the applications of mathematics and the physical sciences to solving real-world engineering problems.

His love for teaching brought him to the academic world. He founded the first Mathematics Resource Center at the University of Alberta: a unit designed specifically to teach high school students the necessary survival skills in mathematics and the physical sciences required for first-year engineering. This led to the Students' Union Gold Key award for outstanding contributions to the University and to the community at large.

Dr. Schiavone lectures regularly to freshman engineering students, high school teachers, and new professors on all aspects of engineering success, in particular, maximizing students' academic performance. He wrote the book *Engineering Success* in order to share with you the *secrets of success in engineering study*: the most effective, tried and tested methods used by the most successful engineering students.

Acknowledgments: I'd like to acknowledge the contributions of: Eric Svendsen, for his encouragement and support; Richard Felder for being such an inspiration; the many students who shared their experiences of first-year engineering—both good and bad; and finally, my wife Linda for her continued support and for giving me Conan.

 **Scott D. James** is a staff lecturer at Kettering University (formerly GMI Engineering & Management Institute) in Flint, Michigan. He is currently pursuing a Ph.D. in Systems Engineering with an emphasis on software engineering and computer-integrated manufacturing. Scott decided on writing textbooks after he found a void in the books that were available. "I really wanted a book that showed how to do things in good detail but in a clear and concise way. Many of the books on the market are full of fluff and force you to dig out the really important facts." Scott decided on teaching as a profession after several years in the computer industry. "I thought that it was really important to know what it was like outside of academia. I wanted to provide students with classes that were up to date and provide the information that is really used and needed."

Acknowledgments: Scott would like to acknowledge his family for the time to work on the text and his students and peers at Kettering who offered helpful critique of the materials that eventually became the book.

 **David C. Kuncicky** is a native Floridian. He earned his Baccalaureate in psychology, Master's in computer science, and Ph.D. in computer science from Florida State University. Dr. Kuncicky is the Director of Computing and Multimedia Services for the FAMU-FSU College of Engineering. He also serves as a faculty member in the Department of Electrical Engineering. He has taught computer science and computer engineering courses for the past 15 years. He has published research in the areas of intelligent hybrid systems and neural networks. He is actively involved in the education of computer and network system administrators and is a leader in the area of technology-based curriculum delivery.

Acknowledgments: Thanks to Steffie and Helen for putting up with my late nights and long weekends at the computer. Thanks also to the helpful and insightful technical reviews by the following people: Jerry Ralya, Kathy Kitto of Western Washington University, Avi Singhal of Arizona State University, and Thomas Hill of the State University of New York at Buffalo. I appreciate the patience of Eric Svendsen and Rose Kernan of Prentice Hall for gently guiding me through this project. Finally, thanks to Dean C.J. Chen for providing continued tutelage and support.

 **Mark Horenstein** is an Associate Professor in the Electrical and Computer Engineering Department at Boston University. He received his Bachelors in Electrical Engineering in 1973 from Massachusetts Institute of Technology, his Masters in Electrical Engineering in 1975

from University of California at Berkeley, and his Ph.D. in Electrical Engineering in 1978 from Massachusetts Institute of Technology. Professor Horenstein's research interests are in applied electrostatics and electromagnetics as well as microelectronics, including sensors, instrumentation, and measurement. His research deals with the simulation, test, and measurement of electromagnetic fields. Some topics include electrostatics in manufacturing processes, electrostatic instrumentation, EOS/ESD control, and electromagnetic wave propagation.

Professor Horenstein designed and developed a class at Boston University, which he now teaches entitled Senior Design Project (ENG SC 466). In this course, the student gets real engineering design experience by working for a virtual company, created by Professor Horenstein, that does real projects for outside companies—almost like an apprenticeship. Once in "the company" (Xebec Technologies), the student is assigned to an engineering team of 3-4 persons. A series of potential customers are recruited, from which the team must accept an engineering project. The team must develop a working prototype deliverable engineering system that serves the need of the customer. More than one team may be assigned to the same project, in which case there is competition for the customer's business.

Acknowledgements: Several individuals contributed to the ideas and concepts presented in Design Principles for Engineers. The concept of the Peak Performance design competition, which forms a cornerstone of the book, originated with Professor James Bethune of Boston University. Professor Bethune has been instrumental in conceiving of and running Peak Performance each year and has been the inspiration behind many of the design concepts associated with it. He also provided helpful information on dimensions and tolerance. Several of the ideas presented in the book, particularly the topics on brainstorming and teamwork, were gleaned from a workshop on engineering design help bi-annually by Professor Charles Lovas of Southern Methodist University. The principles of estimation were derived in part from a freshman engineering problem posed by Professor Thomas Kincaid of Boston University.

I would like to thank my family, Roxanne, Rachel, and Arielle, for giving me the time and space to think about and write this book. I also appreciate Roxanne's inspiration and help in identifying examples of human/machine interfaces.

Dedicated to Roxanne, Rachel, and Arielle

 **Charles B. Fleddermann** is a professor in the Department of Electrical and Computer Engineering at the University of New Mexico in Albuquerque, New Mexico. He is a third generation engineer—his grandfather was a civil engineer and father an aeronautical engineer—so "engineering was in my genetic makeup." The genesis of a book on engineering ethics was in the ABET requirement to incorporate ethics topics into the undergraduate engineering curriculum. "Our department decided to have a one-hour seminar course on engineering ethics, but there was no book suitable for such a course." Other texts were tried the first few times the course was offered, but none of them presented ethical theory, analysis, and problem solving in a readily accessible way. "I wanted to have a text which would be concise, yet would give the student the tools required to solve the ethical problems that they might encounter in their professional lives."

Reviewers

ESource benefited from a wealth of reviewers who on the series from its initial idea stage to its completion. Reviewers read manuscripts and contributed insightful comments that helped the authors write great books. We would like to thank everyone who helped us with this project.

Concept Document
Naeem Abdurrahman- University of Texas, Austin
Grant Baker- University of Alaska, Anchorage
Betty Barr- University of Houston
William Beckwith- Clemson University
Ramzi Bualuan- University of Notre Dame
Dale Calkins- University of Washington
Arthur Clausing- University of Illinois at Urbana-Champaign
John Glover- University of Houston
A.S. Hodel- Auburn University
Denise Jackson- University of Tennessee, Knoxville
Kathleen Kitto- Western Washington University
Terry Kohutek- Texas A&M University
Larry Richards- University of Virginia
Avi Singhal- Arizona State University
Joseph Wujek- University of California, Berkeley
Mandochehr Zoghi- University of Dayton

Books
Stephen Allan- Utah State University
Naeem Abdurrahman - University of Texas Austin
Anil Bajaj- Purdue University
Grant Baker - University of Alaska - Anchorage
Betty Barr - University of Houston

William Beckwith - Clemson University
Haym Benaroya- Rutgers University
Tom Bledsaw- ITT Technical Institute
Tom Bryson- University of Missouri, Rolla
Ramzi Bualuan - University of Notre Dame
Dan Budny- Purdue University
Dale Calkins - University of Washington
Arthur Clausing - University of Illinois
James Devine- University of South Florida
Patrick Fitzhorn - Colorado State University
Dale Elifrits- University of Missouri, Rolla
Frank Gerlitz - Washtenaw College
John Glover - University of Houston
John Graham - University of North Carolina-Charlotte
Malcom Heimer - Florida International University
A.S. Hodel - Auburn University
Vern Johnson- University of Arizona
Kathleen Kitto - Western Washington University
Robert Montgomery- Purdue University
Mark Nagurka- Marquette University
Ramarathnam Narasimhan- University of Miami
Larry Richards - University of Virginia
Marc H. Richman - Brown University
Avi Singhal-Arizona State University
Tim Sykes- Houston Community College
Thomas Hill- SUNY at Buffalo
Michael S. Wells - Tennessee Tech University
Joseph Wujek - University of California - Berkeley
Edward Young- University of South Carolina
Mandochehr Zoghi - University of Dayton

Contents

1

Studying Engineering: The Keys to Success

What does it take to be successful in engineering? The good news is that we *know* the answer to this question: Thousands of engineering students have been doing it for years. As a freshman engineering student, your biggest advantage lies in the fact that many people have already done what you have decided to do, namely, graduate in engineering. To find out what you need to do, you need only draw from the experiences of the many *successful* engineering students that have gone before you. That is what this chapter (and most of this book) is about: the tried *and* tested techniques that will guarantee you success in engineering study.

The most successful engineering students exhibit common key characteristics in their approach to engineering study. The following table lists those characteristics, along with actions typically associated with each:

SECTIONS

- 1.1 Commitment
- 1.2 Application
- 1.3 Strategy
- 1.4 Perseverance
- 1.5 Associations

OBJECTIVES

After reading this chapter you will learn:

- What it takes to be a successful engineering student.
- Techniques that guarantee success in engineering study.
- The study habits of the most successful engineering students.
- The five keys to success in engineering study.

CHARACTERISTIC	ACTIONS
Commitment	*Decide* to be successful. Set appropriate *goals*. Stay focused. Stay *determined* to succeed. Continually remind yourself of the *reasons* you chose engineering.
Application	*Apply* yourself fully to attain your goals. *Work* hard.
Strategy	*Work* smart. Maximize effectiveness. Learn the rules and play the game.
Perseverance	*Don't* give up after the first, second, or third try. *Keep* going. Stay focused on your goals; Use *power thinking!*
Associations	Associate with people that maintain a positive attitude, people that will help you attain your goals. Avoid underachievers and those who do not share your objectives.

In the sections that follow, we discuss each of the preceding characteristics and how they will guide you to success as an engineering student.

1.1 COMMITMENT

When you chose engineering as your career, did you *decide* to be successful, or did you simply *prefer* to be successful? There is a significant difference between the two approaches, particularly when applied to engineering study. When you decide to succeed in engineering:

> There is no alternative: Failure is not an option.

When you simply prefer to succeed in engineering:

> You allow yourself the option of failure.

Each approach has significant consequences for your performance as an engineering student.

When you decide to be successful, you become focused, determined, and committed to success. Graduating in engineering becomes your top priority. You do *not* allow yourself the option of failure. Your mind responds accordingly, allowing you access to the full range of your abilities. This, in turn, maximizes your effectiveness and subsequent performance as an engineering student.

By merely preferring to succeed in engineering, you allow yourself the option of failure: You believe that there is always an alternative, for example, a career in science or business, or perhaps in the "real world." The message your mind gets is that it's okay to fail. Consequently, you become reluctant to apply yourself, you don't try as hard, you lose focus, and you become less determined to succeed. All of this results in less-than-satisfactory performance.

Committed students have no *inner conflict*—they never fight themselves. They know what they want, and they go after it. In doing so, they refuse to lose.

Your level of *commitment* is one of the most important factors in deciding your performance as an engineering student.

> ### Commitment = Deciding to Succeed

Start your path to success by deciding that you will graduate in engineering. Make this your major goal and commit to it. This will equip you with maximum power to achieve that goal. To maintain this commitment, keep in mind the following:

> 1. You chose engineering for definite *reasons*. Stay focused and determined by *reminding* yourself of these reasons frequently.
> 2. Believe in yourself—go for it!

1.2 APPLICATION

Some people find it easy to get good grades in high school without working too hard. This is usually attributed to the fact that they are endowed with some sort of natural academic ability (i.e., they are smart). There is one undeniable fact about engineering study at the university level:

> ### You cannot be successful without hard work!

It has been my experience that many of the so-called smarter freshman engineering students are lulled into a false sense of security, primarily because of their high school experience. They believe that they can carry on as they left off in high school and achieve the same level of success with the same level of application. This belief is always destroyed around midterm time, when grades begin to tumble and they find themselves scrambling to recover. This, of course, is wasteful, counterproductive, extremely stressful, and completely unnecessary.

There are no hard-and-fast rules for how many hours you should spend studying per day, per week, or per semester. Application is more about *productivity* than hours spent. If you spend six hours in the local cafeteria with a group of "study buddies" and devote perhaps 20 percent of this time to doing anything meaningful, then you haven't studied for six hours. So don't fool yourself! The best way to approach engineering study is to accept the fact that you must study as hard as is required of you. In this respect, professors are there to guide you. Relevant and necessary material is presented in lectures, seminars, and laboratories, and assigned as homework. It makes sense to ensure that, at a minimum, all assigned work is:

1. Completed accurately, thoroughly, and in accordance with the requirements.
2. Understood entirely.
3. Recorded clearly for later review—it is always better to *review* than to *relearn.* (See Section 1.3.)

Any additional work, over and above that required and recommended by the professor, is a bonus—like money in the bank!

The very best engineering students are always well aware of the fact that their performance is entirely dependent on how hard they choose to work. At the heart of this is the following belief:

"I'm doing this for me—not for my professor, but for me!"

That's right: Your professors are wonderful people, but at the end of the day, if you perform below your capabilities because you didn't apply yourself, you suffer the consequences, not your professors. It never ceases to amaze me how many freshman engineering students believe they are doing me a favor by attending classes, completing assignments on time, and applying themselves fully to attain their goals. You are the primary beneficiary of all your hard work—not your professors—*you!* Whenever you feel reluctant to sit down and tackle a difficult assignment or attend an early lecture, repeat the following to yourself:

"This is for me! I'm doing this for me!"

The road to an engineering degree is long and hard; if it were easy, the profession would never command the respect it enjoys today. Only very special people are successful in engineering, and never forget that! Take time to remind yourself of the rewards of the profession, but *expect* and *accept* the discipline and hard work required to get there. After all, for the next four years or so, studying engineering is your job, your mission, and your reason for being.

1.3　STRATEGY

Most of us drive a vehicle of some sort. To be able to drive that vehicle on public roads, there are basically two requirements:

1. To learn how to drive.
2. To pass a driving test.

Most of you will agree that knowing how to drive is necessary, but not sufficient, to pass a driving test. A driving test has certain specific requirements. For example, you must be able to execute selected maneuvers in certain ways (as required by an examiner) and demonstrate knowledge on all areas related to the driving experience. When preparing for a driving test, we do not simply spend time driving. That is not enough. Rather, we learn exactly what is required in the test and practice those *specific requirements*. We *rehearse* the actual test over and over again. Let's face it, many excellent drivers would not pass a driving test today without the necessary targeted preparation. Another way to look at this is as that, in deciding to pass a driving test, we:

1. Set the appropriate goal, i.e., to pass the driving test.
2. Discover the rules of the test, i.e., learn exactly what is required to pass the test.
3. Practice the specific requirements of the test.
4. Rehearse the test itself.
5. Pass the test.

Similarly, in studying engineering, there are two basic requirements:

1. To learn all that is required to obtain an engineering degree.
2. To be successful on a (large) number of tests and examinations.

Like it or not, tests and examinations continue to dominate as indicators of student performance at universities. For this reason, they will occupy much of your time and energy over the course of your engineering degree. *Examination technique* (preparing for and writing examinations) is therefore extremely significant in determining your overall performance in engineering. For example, as in the case of driving a car, it is not sufficient to know the course material (even 100%) to demonstrate excellence on a test. The reason is that the latter is a *time-constrained* examination of your abilities. Because of this, there are many additional factors at play that will affect your performance in an examination (e.g., *exam anxiety*, which is discussed in Chapter 6). To deal with these additional factors, effective test preparation must include, among other things, the following two strategies:

1. Find out, learn, and practice the *specific* requirements of the test. (*Target* your review.)
2. Rehearse (and dress rehearse) the test. (Work through sample or past examinations.)

These form the basis of an effective examination technique. We shall return to this topic later, in Chapter 6.

Using effective examination techniques is just one example of working smart. You see, maximizing performance in engineering study is similar to playing a game. As with any other game, there are rules, and the most successful engineering students know these rules intimately and practice them religiously. The following list contains some other examples of working smart (or the rules of the game). These will be discussed in detail in Chapters 3 through 6.

1. *Before a course begins, know what your instructor assumes you know.* One of the main reasons well-qualified students do not do well in engineering is that they approach new courses from the "cold-start" position, not realizing that most courses simply pick up from the end of the prerequisite course (usually taken the previous term). These students spend the first half of the new course relearning the material from the prerequisite course. It has been shown time and time again that a few hour's review of the most important prerequisite skills before the course begins will significantly improve one's

performance. Consequently, ask your instructor to identify the most important prerequisite skills necessary to make an effective start in the (new) course. Make sure those skills are fluent and in working order. You may have forgotten the most important prerequisite techniques (particularly if they were covered the previous semester), so practice with some warm-up exercises if necessary. For example, in calculus, you might practice some factoring, trigonometric identities, or the algebra of functions; in engineering mechanics, you might recall equations of motion and practice drawing some free-body diagrams. Remember, most engineering subjects are cumulative: One part usually depends heavily on a knowledge of the previous part. Consequently, it pays to ensure that your prerequisite actually works. (Your instructor will assume that it does.)

2. *Make the most of lectures and classroom time.* Lectures and classroom time serve to target and highlight relevant material (for both learning and examination purposes). They are used also by the instructor to demonstrate proper procedures (for example, solving problems) and to indicate the required standard. Use lectures and classroom time as your main source of all information relevant to your course. Make sure you know how to take notes efficiently, so as to maximize your effectiveness in the classroom.

3. *Your textbook is a supplement, not a replacement.* Textbooks do not replace lectures or classroom time. There may be far too much material in the textbook, or conversely, there may be insufficient detail on a topic that the instructor deems to be important (and examinable). The instructor will highlight and target the most relevant material. Use the textbook as a supplement (i.e., as a source of examples and practice problems and as a backup).

4. *Write effective solutions to all problems.* Whether you are solving homework, classroom, practice, or examination problems, you should develop an organized problem-solving procedure. The way you arrive at the answer to a problem (i.e., the solution process) is as important as the answer itself. The advantages of a clear, logical, and detailed solution are numerous:

   i. *A clear solution procedure makes it easier to identify errors.* Following a clear, methodical, step-by-step procedure to solve a problem makes it easier to identify mistakes if, for example, you arrive at an incorrect answer. Simply go through each step until you find an error. This is particularly important in answering questions in assignments and examinations.

   ii. *A clear solution procedure makes it easier to score the maximum number of points on an assignment or examination.* In an examination, if you make a mistake that leads to the wrong answer, but have nevertheless demonstrated a clear, logical (and correct) procedure, it is likely that you will receive the majority of the points the problem is worth.

   iii. *Detailed solutions save time when you need to review material.* When you've spent some time solving a problem, it is always worthwhile writing down the details of how you solved the problem. In a few weeks time (e.g., when you need to review for an examination), you will have covered much more material, and it is likely that you will have forgotten how to solve a particular problem. (You may have used some special trick or detail that may have taken a significant amount of time to develop.) Again, it's always easier to review than relearn. A detailed account of the problem-solving process also provides you with a reference to help you solve similar problems when, say, you are practicing for examinations.

iv. *Organize your thoughts and develop an effective problem-solving procedure.* Detail is what distinguishes a *solution* from an *answer.* The former is an account of your thoughts—that is, the problem-solving process (which you should always record)—while the latter is the result of your thoughts. The process is more important than the answer, since it can be applied similarly to other problems. Problem-solving procedures will assist you in all aspects of engineering, (as well as in many aspects of everyday life).

v. *Assignments indicate a required standard.* Homework problems are assigned to allow you to practice *relevant* examples so that you have an idea of the required standard and an indication of what is and what isn't important (for example, in course examinations).

vi. *Practice! Practice! Practice!* We learn most by seeing examples and trying for ourselves. There is no substitute! You cannot swim or drive a car using only the written theory—you must actually perform the deed. Similarly, in engineering, it's easy to read the solution to a problem and believe that you know what to do. However, problem solving requires that you actually perform the solution by yourself, which is entirely different. Practice establishes procedure and allows you to note patterns in solutions so that eventually the procedure becomes automatic (again, as in swimming and driving a car).

vii. *Ask! Ask! Ask!* Never be afraid to ask. Seek out all sources of help, including your instructor. Be polite and professional, but ask as many questions as required. This is a crucial part of the learning process.

viii. *Develop an examination technique.* As mentioned earlier, simply *knowing* the material does not necessarily translate into success on examinations. Remember that in an examination you are subjected to a time constraint. Consequently, you must be as effective as possible by developing an efficient examination technique. For example:

Rehearse and dress rehearse examinations. Old and practice examinations with access to full solutions are an essential part of exam preparation, particularly in eliminating exam anxiety. They afford the possibility of rehearsing the actual examination and therefore reduce the element of surprise.

Write examinations to demonstrate your abilities. When *writing* examinations, explain to the examiner *exactly* what you are doing. Write clear, logical, and detailed solutions to win as much partial credit as possible. On so many occasions, students have returned to my office after an examination to claim that even though a particular one-line answer was incorrect, they knew "how to solve the problem," but didn't feel it necessary to write down the details. You cannot claim that you "know what you are doing" unless you write it down. The examiner requires that you demonstrate your abilities.

1.4 PERSEVERANCE

When you try something new, a certain amount of trial and error is inevitable. Consequently, as you strive towards your major goal of graduating in engineering, you *will* be faced with setbacks, disappointments, and frustrations. Adversity is inherent in engineering study: It's part of the process. How you deal with adversity, however, will determine, to a large extent, whether or not you are successful in engineering. Consider Example 1.1.

EXAMPLE 1.1. John has a big problem with one of the homework questions in Rigid Body Dynamics 250. He has been working on this problem for hours. Not only is it an important assignment problem, but also, it illustrates a key concept that John simply must understand. He consults his professor during office hours. The professor scribbles some equations on John's page and tells John that this should fix his problem. There are other students waiting to see the professor, so John leaves, sits in the hallway, and tries to understand what the professor has told him, but is not convinced. He returns to his professor, who politely and persistently continues to give John the same explanation. John listens respectfully, but keeps coming up with reasons the explanation does not make sense. He once again adjourns to the hallway. Recognizing the urgency of his situation, John returns again to his professor. Politely and professionally, he once more presents the same problem to his professor. By now, John knows the problem intimately, so he knows exactly what to ask. He persists with his counterarguments, but this time he gets what he wants. He leaves the professor's office knowing that he has achieved his immediate goal. John hung in there, pursuing his desired outcome.

The key to John's success was *perseverance.* He had to have the information. This was his immediate goal, and his perseverance allowed him to achieve that goal.

> Perseverance is what drives committed individuals to success.

The previous scenario has been played out many times in my office. The advantages for the student are numerous:

1. The student now knows how to solve the problem.
2. The student now knows that he or she can always come to the professor for help—without feeling intimidated.
3. The professor is impressed with the student's ability to communicate ideas and with the student's enthusiasm for and commitment to his or her work.

Adversity is not what prevents people from achieving their goals; it's what people do in the face of adversity that counts. Successful engineering students persevere whenever they encounter adversity—they don't give up. They hate the idea of not being able to achieve their goals.

Perhaps the most important reason to persevere in engineering study is to develop an effective problem-solving technique, as illustrated by Example 1.2.

EXAMPLE 1.2. Linda encounters a really challenging problem. She consults her notes and looks for something "similar." This allows her to make a start on the solution. Soon, however, the solution path diverges considerably from any of the examples in the class notes and course textbook. She's stuck! What does Linda do next? It's the middle of the evening, so she cannot see her professor until the next day. She can leave it until then, or she can take a break and try again. The latter is what she does. She sits for a while and tries different things, such as, inventing, adjusting, and experimenting with the solution technique. Adversity begins to creep in, but she perseveres. Soon she is very well acquainted with what will work and with what won't work in this particular problem. Then, all of a sudden it hits her! Bingo! She solves the problem.

Have you ever had such an experience? Many students have told me of similar experiences and compared them to air travel. That is, to solve the problem, they had to move up a level of thinking. As in air travel, moving up a level meant that it was necessary to go through the inevitable turbulence (adversity). Once they had fastened their safety belts (analogous to perseverance) and passed to the next level, they entered into the *power-thinking* zone, which enabled them to solve the problem. Entering the power-thinking zone may not be sufficient (i.e., you may have to ask for outside help eventually), but it certainly is necessary to learn how to solve the problem yourself.

We summarize effective problem solving in the following three stages:

Stage 1   Acquaint yourself with the details of the problem, define the goal, and decide that nothing will stop you from achieving that goal.

Stage 2   Try some obvious solutions. Use class notes or the course textbook. This is where adversity begins to creep in. You become frustrated with each setback.

Stage 3   Here, you are well acquainted with all details of the problem. You know more or less what doesn't work, and you have narrowed down the search to a few alternatives. You may seek some help, but you are highly focused nonetheless. This stage allows for power, *or* deep thinking. The problem is solved here.

Perseverance is what allows you to progress to Stage 3. In addition, the following are true of perseverance:

1. *Ingenuity increases with perseverance.* As you move from stage to stage, you try smarter or more informed solutions, you really hone your problem-solving skills, and you *learn* from the experience.
2. Perseverance is necessary to *warm up* your thinking so that you may enter the *success zone*, i.e., so that you may proceed from Stage 1 to Stage 3.
3. Perseverance allows you to break through into *power thinking*.

1.5 ASSOCIATIONS

Have you ever noticed how you tend to pick up the habits of the people with whom you associate the most? For example, you may find yourself:

Using the same words or phrases.	Doing the same things.
Believing the same things.	Finding the same things distasteful.
Enjoying the same things.	Developing the same attitudes.

You may do all of these things just by association.

If you are committed to achieving your goal of graduating in engineering, your choice of friends and colleagues is extremely significant when it comes to progressing towards that goal. Imagine for a moment that your best friend:

1. Always missed the 8 o'clock class in favor of an extra snooze.
2. Almost never handed in assignments on time.
3. Was part of a study group for the sole reason of copying solutions to assignments.
4. Made a habit of leaving everything to the last minute in favor of more important things such as movies, football games, nightclubs, and parties.
5. Retained certain high school attitudes, including the belief that it was *cool* to fail.
6. Talked or read the newspaper during lectures.

At best, a friend like this will distract you sufficiently to ensure that you never perform to your full potential. At worst, you will inherit similar characteristics and begin the downward spiral to failure!

In studying engineering at the university level, it is *essential* that you align yourself with people who:

Share your objectives.	Strive for excellence.
Will stretch you and *push* you towards your goals.	Are positive.
Share your attitudes on studying engineering.	Are suitable role models—share your hopes and dreams.

Your engineering education represents a considerable investment in time, money, and effort. This is your *job* for the next four years or so, and you should do everything possible to maximize the effectiveness of your learning environment.

Begin by making a list of all the people that are significant in your life. Ask how they contribute to or detract from your goals, the most important of which is graduating in engineering. Then, systematically get the negative people out of your life. It takes a lot of time and energy to achieve your goals, you do not need the added burden of people who will slow you down and prevent you from realizing your dreams. Fill your life with positive, energizing people, people whom you admire and who are what you want to be.

Problems

1. Why did you choose to study engineering? Write down your top 10 reasons, the first being the most important reason.
2. List five benefits you expect to obtain when you graduate with an engineering degree.

3. List five reasons *not* to choose engineering.

4. Make a list of goals you hope to achieve:
   a. today.
   b. this week.
   c. this semester.
   d. this year.
   e. by the time you graduate with your bachelor-of-science degree in engineering.
   f. in your lifetime.

5. List all the New Year's resolutions you have *ever* made. Next to each one, say whether or not you were successful. Beside those identified as successful, write down five reasons you were successful. Why were you successful in some resolutions, but not in others?

6. List five occasions when you decided you would do something and *successfully* carried out your intentions. List five occasions when you decided you would do something, but did *not* follow through with your intentions. List the reasons why you were successful in some cases, but not in others. What did you do that was different?

 How many hours a week did you study in your last year of high school? How many hours a week do you expect to study at the university? List five reasons you need to study harder at the university.

8. Consider the following table, which describes a typical course load from the first year of an undergraduate engineering student:

TERM 1	TERM 2
Chemistry I	Chemistry II
Introduction to the Engineering Profession I	Computer Programming
Engineering Mechanics	Introduction to the Engineering Profession II
Calculus I	Engineering Physics
Physics I	Calculus II
Elective	Linear Algebra

Make a similar table of your own first-year courses. Next to each course, write down the number of hours per week you would need to devote to each course to get an excellent grade. You should take into account the following:

> Lecture hours
> Laboratory hours
> Seminar hours
> Time required to review lecture notes
> Time required to complete a weekly assignment
> Time required to study new concepts and fix ideas

Use the preceding information to compile a detailed weekly study schedule.

9. List the (high school) courses that are a prerequisite for each of your (university) engineering courses. List the most important skills from each prerequisite course that you think your professor will assume on day one of each university course. For example,

Course	Prerequisite Skill
Calculus I	Factoring, functions, trigonometric identities
Engineering Mechanics	Free-body diagrams, concepts of force, mass, etc.

Confirm your answers with the corresponding professors.

10. Make a list of all the *nonacademic* examinations you have taken in the last five years—for example, driving tests, stage performances, athletic competitions, fitness tests, tests related to extracurricular activities and sports. In how many cases did you undertake a *dress rehearsal* before the actual test itself? How did you practice to ensure that you would be successful on each test?

11. Using the information in Section 1.3, list five ways in which you could improve your performance in engineering study.

12. Do you believe that you will *bother* your professor if you go to his or her office to ask questions? Why do you believe this? List five duties you expect a professor to perform in a typical day.

13. List five outcomes in your life that have been achieved through perseverance and determination. For each outcome, list the driving factors that kept you focused and determined until you reached your goal.

14. What's more important to you: the *answer* to a problem or the *solution*? Why? List the differences between the two. Illustrate the differences by providing the solution and the answer to the following problem:

$$\text{Differentiate the function } f(x) = x^2 \sin x$$

15. List 10 skills you believe to be essential for success in undergraduate engineering study. Rate these skills, beginning with the most important and ending with the least important. Ask a professor to compile his or her own list, and compare it with yours. Revise your list by identifying any items you feel should have been added.

16. Using the final list in Problem 15, identify your strongest and weakest skills, and decide how you will improve on the skills that need attention.

17. Make a list of all the significant people in your everyday life. Decide which of these people help you grow emotionally and intellectually, push you to achieve more, encourage you, and fill you with positive energy. Decide also which people hold you back and divert you from your main objectives. Resolve to spend more time with the former and less, if any, with the latter.

2

Introduction to Engineering and Engineering Study

"How much do you know about engineering? Why did you choose to study engineering?

What reasons lead you to believe that you are *ready* and *equipped* to study engineering?

What are the main differences between studying at a university and studying in high school?

What new *success skills* do you need to succeed in engineering study?

Can you write down 10 answers to each question I have asked you? Go ahead and try."

This is often how I begin my lecture to freshman engineering students enrolled in an *introductory engineering* class. After a little thought, most of them realize just how little they know about this subject called *engineering* and (often despite excellent high school averages) how ill equipped they are to *study engineering*.

In this chapter, we address both issues. First, we ask the following questions:

1. What is engineering?

2. What do engineers do?

3. Why choose to study engineering?

SECTIONS

- 2.1 What Is Engineering?
- 2.2 What Do Engineers Do?
- 2.3 Why Choose to Study Engineering?
- 2.4 Equipping Yourself for Engineering Study
- 2.5 Cooperative Education Programs (Co-ops) and Internships

OBJECTIVES

In this chapter you will:

- Be introduced to the engineering profession.
- Learn about different engineering disciplines and the areas of specialization within these disciplines.
- Learn about the different jobs that engineers do and the industries that employ them.
- Discover the rewards and opportunities offered by a career in engineering.
- Understand the main differences between studying in college or university and studying in high school.
- Learn new *skills* required to succeed in engineering study.
- Find out about cooperative education programs and internships.

13

The answers to these questions are not only interesting and informative, but will help keep you motivated along the long, hard road to an engineering degree. Ability and hard work might get you through the initial stages, but after that, you must have a driving force, something that will sustain you through the hard times. You must develop a powerful motivation. The best way to do this is to learn as much as possible about the rewards of an engineering degree. Perhaps write them out and pin them on your wall or paste them inside your calculus book. Keep them close at hand. They will keep you determined and strong. This is exactly what the most successful engineering students do; they remain focused by keeping in mind the reasons they chose engineering and the rewards associated with entering the engineering profession. Make it a priority to keep learning about engineering, so that you will become aware of all the opportunities and rewards as they arise throughout your course of study. This will fuel your motivation and your desire to succeed. The more important it becomes for you to graduate, the more likely you are to do so.

In Section 2.4, we address the question, "Are you prepared and equipped for engineering study?" In doing so, we examine the study skills required to succeed in the university environment. For many students, the university is the next logical step after high school, the next academic challenge. Consequently, they expect their freshman year in engineering to be much like another year of high school—which, of course, it isn't. In engineering, such an exception often manifests itself in unacceptably high first-year attrition rates. We address this issue by focusing on what you need to do to ensure the best possible start to earning your engineering degree. Essentially, you must develop the necessary:

Work strategies

Study strategies

Attitudes

Communication skills

Ability to work as part of a team

Time management skills

2.1 WHAT IS ENGINEERING?

What does *engineering* mean to you? Here are a few suggestions offered by freshman engineering students:

"A subject that reflects our understanding of things around us"

"The application of scientific knowledge to solve practical problems"

"The bridge between pure science and practical application"

"The application of scientific principles to provide goods to satisfy human needs"

"Creative problem solving"

"The use of technology to perform tasks"

"The study of how to build things"

"The study of how things work and how we can make them work better"

"Creating, designing, testing, and improving systems"

"A scholarly, yet practical, study of the physical applications of human beings' technology combined with nature's laws"

"A profession by which you utilize mathematical, scientific, and physical knowledge for the betterment of humankind"

"Applying math and science to life"

"The application of the simplest and least costly method to solving a problem"

"Being creative and facing new challenges every day"

In fact, engineering is all these things. The activities included in its definition are complex and varied. It is an ever-expanding subject area. To me, engineering is

> **The practical application of math and science to create, design, test, improve, and develop knowledge, research, money, business, economics, and technology.**

This is why engineering is such a challenging and demanding field of study: It involves areas of expertise that continue to evolve independently, yet are required to perform together as part of the engineering process. Thus, an engineer must be expert in many areas, must know how to communicate knowledge between those areas, and must apply that knowledge to create, design, study, research, and invent all kinds of things. It is not uncommon for engineers to begin their careers as mathematicians, applied scientists, or even economists as I did.

The early stages of an engineering degree must be broad based. The initial emphasis on mathematics, science, technology, and language arts is no accident: These are the building blocks for what will follow in the later years of specialization within engineering.

> *Engineering is a process that applies mathematics and physical science to the design and manufacture of a product or service for the benefit of society.* This process is illustrated in the following diagram.

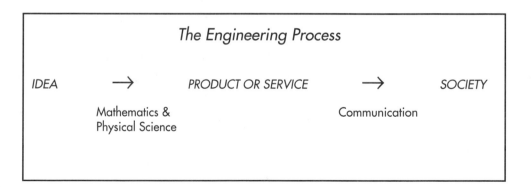

2.2 WHAT DO ENGINEERS DO?

Engineers are classified largely according to their areas of specialization. Traditionally, these areas have been:

Mechanical engineering

Electrical engineering

Civil engineering

Chemical engineering

Industrial engineering

More recently, the following areas have emerged as separate engineering disciplines:

Materials engineering

Computer engineering

Other areas of further specialization arising from advances in technology and the growth of related areas include the following:

Biomedical engineering	Mining and petroleum engineering
Environmental engineering	Agriculture and biosystems engineering
Aerospace engineering	Manufacturing engineering
Nuclear engineering	Ocean engineering/naval architecture

In what follows, we summarize the main features of each of these engineering disciplines and list the different occupations associated with each area of specialization.

2.2.1 Mechanical Engineering

Mechanical engineering is concerned with the analysis, design, and development of structures, machines, devices, and mechanical systems. Consequently, mechanical engineers work in areas related to:

The design and development of machinery and devices

The analysis of mechanical systems and the vibrations of structures

The design and development of manufacturing processes and energy conversion systems

The design of heating, ventilation, and air-conditioning systems

Mechanical engineering is perhaps the broadest of the engineering disciplines, covering a large number of technical fields broadly classified into the following four areas:

1. Solid mechanics
2. Fluid mechanics
3. Thermodynamics
4. Mechanical design

Solid mechanics Engineers working in solid mechanics are concerned with analyzing the behavior of solid bodies subjected to stresses and external loads. For example, in the design of a new bridge, it is necessary to predict how the bridge will react to high winds and the additional stresses caused by vibrations. Mathematical models are used to predict the behavior of the solid body. This information is then utilized in the design and manufacturing processes. For example, in the design of structures and mechanical components in engines, a stress analysis allows the designer to select materials and configurations for optimum performance. Applied mathematicians and physical scientists are equally at home in this area of mechanical engineering, allowing for interdisciplinary research and development.

Mechanical engineers specializing in solid mechanics find employment in many industries, including those related to bioengineering; tribology; aerospace; the design and manufacture of advanced materials, textiles, and composite materials (e.g., ceramics and fiber-reinforced materials); the design and manufacture of engines; acoustics; management; nondestructive evaluation; the design and manufacture of pressure vessels; and dynamical systems and control.

Fluid mechanics Fluid mechanics is concerned with the behavior of *liquids and gases* and the design and development of machinery, such as pumps, pipes, fans, and turbines, responsible for that behavior. For example, in the aerodynamics industry, mechanical engineers specializing in fluid mechanics apply their expertise to the flow of air around aerofoils to design and develop many of the major components of aircraft, including jet engines.

Mechanical engineers specializing in fluid mechanics find employment in many industries—meteorology, oceanography, acoustics and noise control, fluid power systems, aerodynamics, and the design and manufacture of pressure vessels, to name just a few.

Thermodynamics   Thermodynamics is concerned with the conversion of one form of energy to another, as, for example, in the production of electricity. Here, energy from the combustion of fuels such as coal, oil or natural gas is converted to mechanical energy, which drives a generator. Mechanical engineers in this area apply their expertise to the design and development of devices such as power plants, engines, heat exchangers, and cold-storage facilities.

Specific examples of employment in the area of thermodynamics are in: environmental control; heating, ventilation and air-conditioning systems; alternative fuel sources; pollution control; and solar energy, among others.

Mechanical design   Mechanical design can be described as a process that translates an idea, demand, or identified need into a working prototype of a product or service. Accordingly, mechanical design engineers are involved in many different aspects of the engineering process. For example, they must gather all available information on the proposed product or service, perform the necessary analysis, optimization, and evaluation, and, finally, produce a working prototype. The availability of many different designs (solutions) for a given set of requirements (problems) makes the task of finding the best design extremely challenging, yet this is one of the most rewarding aspects of mechanical engineering. Design engineers find employment in any area that develops, improves, and manufactures products or services, for example, everything from the manufacture and development of automobiles and aircraft to the production of office equipment (e.g., photocopiers, fax machines, pens, pencils, etc.), computers, and kitchen and home appliances.

2.2.2 Electrical Engineering

Electrical engineering embodies the study of all things electrical—in particular, electrical devices, electrical systems, and electrical energy. Consider for a moment the extent to which electrical phenomena are present in our everyday lives, as computers, cars, televisions, stereos, machines used in manufacturing, automatic banking machines, and communications devices (cellular phones, fax machines, etc.). Soon you will begin to see the importance of the work of electrical engineers. As a graduating electrical engineer, you will find no shortage of employment opportunities—in every sector of the economy.

There are five major specialities in electrical engineering:

1. Electric power engineering
2. Communications
3. Control systems engineering
4. Digital systems engineering
5. Electronics

Electric Power Engineering   Electric power engineers, or simply power engineers, design, develop, and maintain systems and devices for the generation, transmission, and distribution of electric power. A strong background in mathematics, circuit analysis, control systems, electric machines, and computing allows the power engineer to be involved in all aspects of the planning, transmission, utilization, and control of electrical energy.

Communications   Communications engineering is concerned with the transmission of information using wires, coaxial cable, fiber optics, or radio. Communications engineers must be familiar with various methods of transmitting, routing, and receiving both analog and digital signals, as well as methods for processing these signals. A strong background in network analysis and statistical analysis is required for the design of complete communication systems, while an in-depth knowledge of digital and analog circuit design, signal processing, and electronics is essential for the design of individual elements making up such systems. Recent technological advances in fiber optics and lasers, together with the revolution in personal communications (e.g., cellular phones, the Internet, etc.) mean that there are extensive opportunities in this area of engineering.

Control Systems Engineering   Control systems engineering is concerned with the design and development of machines or systems that control automated processes. The control of physical systems is very much an interdisciplinary area involving many engineering specializations; consequently the control systems engineer will find employment in many areas, including robotics, manufacturing, the aerospace industry, offshore oil and gas extraction, power systems, and the manufacture of automobiles and household appliances.

Digital Systems Engineering   Digital systems engineers draw on expertise in the areas of digital system design and digital electronics to design hardware for a broad range of applications, including digital signal processing, communications, computers, and instrumentation. For example, a digital systems engineer might design an electronic device to measure, process, store, and transmit data. The need for this kind of expertise is widespread in such areas as commercial aircraft, oil exploration, telecommunications, and banking, allowing for extensive employment opportunities in these industries.

Electronics   Electronics is an area of electrical engineering that is concerned with the design and development of electronic devices and electrical circuits for the production, detection, and control of electrical signals. Electronic engineers use knowledge in the areas of solid-state devices, integrated circuits, semiconductors, and computer-aided design to design and analyze systems used in a wide variety of applications from microwave systems to instrumentation. Consider the impact that recent advances in electronics have had on your everyday life. For example, personal computers, cars and trucks, household appliances, and personal stereo equipment are commonplace in many households.

2.2.3 Civil Engineering

Civil engineers plan, design, construct, operate, and maintain many of the structures and facilities around us, for example, airports, buildings, bridges, harbors, highways, transit systems, offshore drilling platforms, waste collection structures, water supply facilities, and other public works. Civil engineering offers vast employment opportunities in a wide variety of technical fields ranging from planning, design, and construction to research, teaching, sales, and management. There are seven major specialities within civil engineering:

1. Construction engineering
2. Environmental engineering
3. Geotechnical engineering
4. Structural engineering
5. Surveying
6. Transportation engineering
7. Water resources engineering

Construction Engineering Construction engineers combine engineering and management skills to plan and complete projects designed by architects and consulting engineers. Such projects might include the construction of a bridge, building, or shopping mall. The construction engineer will be responsible for the entire project, from start to finish. He or she will lead a team of financial planners, technicians, tradespeople, and professional engineers from many specialities to ensure that the project is completed on time, within budget, and in accordance with the designer's specifications. The construction engineer will be responsible also for the choice of construction methods, materials, and equipment. He or she also ensures that the work is performed safely.

Environmental Engineering Environmental engineers provide technical solutions to environmental problems, such as those associated with the prevention and control of pollution in water and air and on land. In the latter regard, the environmental engineer will work in areas related to the provision of safe drinking water; disposal and recycling of waste (including hazardous waste); provision of municipal services such as sewer, water mains, and garbage disposal; air pollution control; reclamation of industrial land; and cleanups at sea (e.g., from oil spills).

Geotechnical Engineering This branch of civil engineering is concerned with the analysis of the properties of soils and rock that support and influence the behavior of structures, pavements, and underground facilities. Geotechnical engineers evaluate the potential settlement of buildings, the stability of slopes and fills, the potential for groundwater seepage, the potential for landslides, and the effects of earthquakes. They work closely with structural engineers in the design and construction of dams, tunnels, and building foundations.

Structural Engineering Structural engineers are responsible for the planning and design of all types of structures, from bridges and containment facilities to buildings, high towers, and drilling platforms. They analyze the forces that each structure must withstand, select the appropriate construction materials, and configure all members and connections to produce a safe, stable, effective, and economical structure.

Surveying A surveyor makes precise measurements to obtain reliable information, which is used to locate and design engineering projects. For example, surveyors' maps give accurate information for the construction of tunnels, highways, and dams. Modern-day surveying makes use of the most up-to-date technology, including satellites, aerial and terrestrial photogrammetry, and computers.

Transportation Engineering Civil engineers working in the area of transportation plan, design, construct, and manage all types of transportation facilities, including streets and highways, transit systems, airports, railroads, ports, and harbors. They are also involved in the provision of safe methods for transporting goods, such as oil and gas and hazardous materials.

Water Resources Engineering   This area of specialization is concerned with the control and use of water, including flood control and protection, water distribution systems, wastewater collection systems, irrigation, drainage, hydroelectric power, harbor and river development, and road and pipeline river crossings. Water resources engineers draw from a strong background in hydraulics, hydrology, fluid mechanics, mathematics, and computer analysis.

2.2.4 Chemical Engineering

Many of the products we use in our everyday lives (e.g., food products, building materials, plastics, oil, gas, and electricity) are conceived and developed by scientists working in laboratories. Chemical engineers are responsible for translating these *small-scale* successes into *large-scale* commercial realities for the benefit of society as a whole. Chemical engineers design, build, maintain, and develop the complex systems required to convert the laboratory experiment into an industrial operation capable of large-scale production. They apply principles from chemistry, physics, and mathematics in order to understand and overcome problems associated with, for example, heating or cooling large quantities of materials or moving these same materials from one section of the plant to another. Equipment used inside the plant is designed and built by the chemical engineer to withstand the necessary high temperatures and pressures, as well as the inevitable corrosion and wear.

Graduates in chemical engineering find employment in many industries, including the chemical, petrochemical, food-processing, forestry, and pharmaceutical industries.

There are six major specialities within chemical engineering:

1. Polymer engineering
2. Biotechnology
3. Process control engineering
4. Environmental engineering
5. Engineering management
6. Oil and natural gas

Polymer Engineering   Polymer engineering is an area of materials engineering in which advanced materials composed of chainlike polymer molecules are designed. Examples of such materials are plastics, rubbers, fibres, films, and composites.

Biotechnology   New biotechnology industries produce a range of products for medical, agricultural, food, and chemical applications. Chemical engineers apply principles from the life sciences—in particular, molecular biology—to design, develop, and operate complex processes for the manufacture of these new products. For example, chemical engineers develop processes for the economic production of fertilizers. One use of these fertilizers would be eliminating famine in developing nations.

Process Control Engineering   Process control is concerned with the design and development of control systems to maintain the efficient operation of large-scale industrial processes. Process control is necessary to account for changes in any of the process parameters, such as a change in raw materials. Control systems range from schemes that use simple instruments to sophisticated digital computer systems.

Environmental Engineering   Many chemical engineers are involved in designing and developing technical and economically feasible solutions to environmental pollution problems in an effort to protect and improve our environment.

Engineering Management   Chemical engineers often assume the responsibilities of technical managers by becoming involved in the design and implementation of optimal operating conditions inside a complex, large-scale industrial plant. In this case, both technical and managerial skills are required for the smooth operation of the process.

Oil and Natural Gas   Chemical engineers are frequently employed in industries that produce, process, and refine natural gas and petroleum and in those industries that manufacture petroleum products, including soaps and cosmetics.

2.2.5 Industrial Engineering

Industrial engineering is concerned with *efficiency*, or, more precisely, how to design, organize, implement, and operate the basic factors of production (materials, equipment, people, information, and energy) in the most efficient manner possible. Because of this orientation, industrial engineers are concerned with related issues, such as optimum performance, reliability, cost efficiency, quality control, plant design, and management of human resources. Opportunities for industrial engineers are extensive They may find employment in many areas, from the financial sector (banks, insurance companies, etc.) to the medical profession and the public sector.

The more *modern* engineering disciplines are materials engineering and computer engineering, discussed briefly in the next two sections.

2.2.6 Materials Engineering

Materials engineering has evolved as a separate discipline mainly because of the production and engineering applications of new, advanced materials (polymers, ceramics, composites, and electronic materials). Materials engineers are involved with materials production, materials processing, and materials application and design. They find employment in all sectors of the economy that involve the materials cycle, including raw materials processing (steel production, aluminum smelting, and mineral processing), the manufacturing sector (the aerospace, automotive, electronics, and petrochemical industries), and the service sector (tribology, fracture mechanics, failure analysis, and recycling).

2.2.7 Computer Engineering

The use of computers in today's society is pervasive. Computer engineering, once considered part of electrical engineering, has evolved as a separate discipline in response to the demand for, and widespread use of, computers. Computer engineers utilize knowledge from both electrical engineering and computer science to design and implement computer systems in which the hardware and software components are both intimately connected with and critical to the success of the design. Examples of such systems can be found in control processes and in the communications industry. Graduates find work in areas ranging from software design and systems analysis to electronics design and the design of specialized electronic devices. Examples of specific areas of employment for computer engineers are industries related to the design and production of automobiles, biomedical equipment, communications equipment, and chemical plant control systems.

2.2.8 Areas of Further Specialization

Biomedical Engineering   Biological engineering is concerned with the application of engineering concepts and technologies to solve biological and medical problems. Biological engineers work with biologists and medical doctors as part of interdisciplinary teams to design and develop new technologies such as artificial organs, prosthetic devices, and medical equipment. They are employed in both the private and public sectors, from small consulting companies to pharmaceutical companies, medical device manufacturers, and government agencies.

Environmental Engineering   Environmental engineering deals with issues concerning the environment, such as urban, regional, and global air quality; water supply and water quality control; hazardous waste treatment; global environmental change; the maintenance of stable ecosystems; prevention and control of air pollution; wastewater management; and hazardous waste management. Environmental engineers work in both the private and public sectors. For example, oil companies will employ environmental engineers to ensure that oil and gas resources are extracted with minimum impact on the environment. On the other hand, local municipalities will employ environmental engineers to oversee the management of drinking water and waste disposal systems.

Aerospace Engineering   Aerospace engineers design, develop, and implement new and existing technologies in both civil and military aviation, including the design and development of all types of commercial and military aircraft, such as vehicles used for space exploration. In addition, aerospace engineering is concerned with the development of systems that support the safe and efficient operation of all types of aircraft. For example, aerospace engineers are involved in control and guidance systems, information systems, and instruments used for navigation. There are many specialities within aerospace engineering. For instance, the more theoretically inclined may specialize in aerodynamics, in which mathematical models are used to predict the behavior of different materials and new designs in the construction of aircraft. On the other hand, the more practically minded may be involved in the manufacture of commercial and military aircraft. Other areas of specialization include acoustics, rocket technology, computational fluid dynamics, and thermodynamics.

Nuclear Engineering   Nuclear engineers deal with all aspects of nuclear power, from the design, development, and implementation of different forms of nuclear power (e.g., nuclear power plants for the generation of electricity and nuclear engines in submarines and spacecraft) to the handling and safe disposal of nuclear fuels used in the commercial and military sectors of the economy. Radioactive materials can be found in hospitals, medical clinics, and laboratories, as well as in military establishments, where they are used in the construction of advanced weapon systems.

Mining Engineering and Petroleum Engineering   Mining engineering is concerned with the extraction of minerals and hydrocarbons from the earth and the processing of these minerals in preparation for further use. Mining engineering is a broad field that includes specialities such as methods for estimating ore reserves, geostatistics, geology, underground and surface mine design, the design and development of mining equipment, rock and soil mechanics, and mineral processing. Petroleum engineering is also a speciality within mining engineering in which the primary focus is on oil and gas. Mining engineers work all over the world, in both developed and developing nations. They are employed by a range of companies in both the private and public sectors, including mining companies, oil companies, government and industry research facilities, and even financial institutions, which invest heavily in these types of industries. A mining or petroleum engineer might be employed to design mines and mining equipment, supervise and manage mining operations, manage and design blasting operations, or research occupational health and safety in the workplace.

Agriculture and Biosystems Engineering   Agricultural engineering is concerned with efficient food production while maintaining or improving the environmental quality of the agro-ecosystem. Agricultural engineers deal with the design, development, construction, and operation of systems for food production, storage, handling, and processing. They are trained in biological and agricultural sciences, and they use their skills not only to improve the efficiency of food production, but also to decrease or eliminate environmental hazards and reduce the consumption and waste of natural resources. The role of the modern-day agricultural engineer has changed dramatically with the ever-expanding field of biotechnology. For example, agricultural engineers now provide specialized services in areas such as the genetic manipulation of plants and animals, the development and implementation of environmentally friendly pesticides, and the production of enzymes in the enhancement of food processing. In addition, they continue to provide expertise in traditional areas, including farm equipment and structures such as tractors, barns, drainage systems, harvesters, and processing units. Agriculture and biosystems engineers are employed by consulting companies, government agencies, small businesses, and academic institutions all over the world.

Manufacturing Engineering   Manufacturing engineering is concerned with the design, development, and implementation of all aspects of manufacturing operations, from product, equipment, and inventory to quality management, on-time delivery, capacity, and manufacturing cost. Because of this concern, manufacturing engineers are involved in many of the things that we use in our everyday lives, from the clothes we wear to the vehicles we drive. Both the private and public sectors employ manufacturing engineers in any operation that involves the manufacture of a product or service.

Ocean Engineering and Naval Architecture   Ocean engineering and naval architecture are concerned with the application of ocean science and engineering design to the ocean environment. Ocean engineering differs from conventional land-based engineering mainly as a result of additional factors that must be taken into account in designing and manufacturing for the ocean environment. Examples of such factors are wave motion, currents, significant temperature variations, and chemical and biological factors. Ocean engineers and naval architects design offshore drilling platforms, harbors, and the corresponding equipment required for their maintenance. These engineers also apply their engineering expertise to the design and development of ships and other water-based vessels, as well as to help solve the problems related to beach erosion. Ocean engineers and naval architects find employment in both the private and the public sectors—for example, in consulting firms, environmental agencies, finance, shipbuilding, and government, and on offshore drilling platforms.

2.2.9 Engineering Job Classifications

Engineers are classified not only by an *area* of specialization, but also by the *type* of engineering they perform within that area. For example, a mechanical engineer, could also be classified as an analyst, an experimentalist, a design engineer, or a research engineer. Your own particular strengths and preferences will determine which of these you choose to be.

Analytical Engineer   The analytical engineer is concerned mainly with the mathematical modeling and analysis of engineering problems. It is often far too expensive to build and test real, *physical* prototypes (e.g., an aircraft wing). Instead, analysts often build and analyze cheaper *mathematical models* for the preliminary investigations of the viability of a new idea, design, or concept. Analytical engineers have a strong background in applied mathematics, the physical sciences, engineering science, and computer applications.

Experimental Engineer   In contrast to the analytical engineer, the experimental engineer is concerned mainly with physical prototypes, and he or she will use actual *hands-*

on experience to evaluate a particular new idea. The experimental engineer is more comfortable in the laboratory or out in the field than in the office.

Design Engineer   The design engineer is involved in all aspects of the design process. He or she will use ideas and relevant information to produce a detailed plan, or *design,* from which an actual product can be assembled. The key word in design engineering is *detail.* Often, plans and specifications are read by a nonexpert who will require precise instructions. Also, the most technically brilliant design is not necessarily the best one, because of cost considerations or the unavailability of materials. Hence, in achieving the optimum design, the design engineer must also consider other factors, such as efficiency, cost control, and ease of production.

Research Engineer   The research engineer is concerned with the development of *new* products, designs, and processes. The research is usually *applied,* in the sense that it is directed towards a fixed goal rather than being open ended, as in the case of curiosity-driven research. Most industries employ at least a small number of research engineers, if only to ensure that their products and services remain at the leading edge of technology. Research skills include the ability to find and assimilate *relevant* information and use that information in conjunction with engineering skills to develop new products or improve existing ones.

Test Engineer   Test engineers *test* new and existing (quality control) products and processes to see whether they comply with the required design specifications. For example, a test engineer might test a new braking system on a car or a new control system to see if it is reliable.

Consulting Engineer   The consulting engineer works as an independent professional, selling his expertise to clients, usually on a contract basis. Consulting engineers are required to be registered as professional engineers before undertaking any form of consulting. This assures clients of the engineer's expertise and professionalism. Typical services offered by consulting engineers include design, analysis, investigations for insurance companies, professional witness services (e.g., testifying in vehicle accident cases), research and development, management, construction, and environmental research. In fact, professional engineers can be found working in every engineering speciality.

Engineering Management   Engineering managers are engineers (from any of the engineering specializations) who *combine* engineering skills with managerial abilities to *direct* resources towards the efficient production of goods and services. They can manage whole projects or different parts of projects. The chief objective of an engineering manager is no different from that of any other manager: to see that a project or task is completed according to the required specifications, on time, and within budget. Engineers usually move into this area through promotion, having demonstrated solid leadership and technical ability.

Engineering Professor   Engineering professors are employed by colleges and universities as teachers and researchers. A Ph.D. degree is the minimum qualification required to enter an engineering faculty. As with professional engineers, engineering professors are classified according to their particular speciality. For example, there are professors of mechanical engineering, of electrical engineering, and so on. On the instruction side, professors develop and teach different engineering courses at various levels (junior undergraduate, senior undergraduate, and graduate), improve and update curricula, and supervise graduate students. Their research activities include discovering and developing new information, publishing new findings in scientific and engineering journals or books, soliciting research funds from both private and public sources, presenting information at conferences, and collaborating with scholars from other institutions. Engineering professors also undertake what is known as *community service* by participating in professional societies, consulting, and serving on education or various other government boards.

2.3 WHY CHOOSE TO STUDY ENGINEERING?

What are your reasons for choosing to study engineering? Write down as many as you can. You'll find it an interesting and rewarding exercise. The following are some of the reasons offered by my freshman engineering students:

"It's really interesting."

"Engineering is practical."

"[It's] useful in life."

"I get to make a difference in society."

"To please my parents."

"I like to think."

"The money."

"[The] variety of challenging and exciting problems."

"[You] get to travel and meet lots of interesting people."

"If you can finish an engineering degree, you can do anything."

"Prestige."

"Respect."

"[There's a] wide range of careers available within the engineering profession."

"My parents told me I had to because it would get me a stable, secure job."

"To learn how to make practical and significant decisions."

"To train my mind and to give me confidence."

"Job security."

"Engineering is an international subject. I can work anywhere in the world."

"I really enjoy applying math to physical, practical problems."

"Status in society."

"[It has a] well-defined career path and options."

"[There are] so many different types of study under the same umbrella."

"I have a knack for making things and understanding how things work."

"[You can] work on a team with other engineers."

"It's a dynamic subject—new fields of engineering are created every day!"

"My mom and dad are both engineers—it seemed the natural thing for me to do."

"I love applying math and physics."

"[It gives me] a chance for personal development."

"I want to do something more than my parents did."

"I enjoy being creative."

How many reasons did you find? Twenty. Thirty. Forty? The reason for this variety of reasons is that the rewards, benefits, and opportunities offered by an engineering education are vast. A career in the engineering profession is just one of them. Priorities may be different (e.g., some may prefer the high standard of living that comes with an engineering career, while others might enjoy the challenges and the variety that are a part of studying engineering), but most engineering students agree on the following list of major reasons they chose to study engineering (in no particular order):

1. Engineering is really interesting and enjoyable.
2. Engineering offers personal and intellectual development.
3. Engineering is challenging—not just now and again, but every day.
4. Engineering offers a variety of career opportunities.
5. Engineering offers financial security.
6. Engineering is being able to apply math and science to solve real-world problems and understand how things work.
7. The engineering profession is respected and prestigious.
8. Engineering offers the opportunity to be creative.
9. Engineering offers the opportunity to do something good for society.

Your own personal preferences will dictate how you rank these different items. Most students agree, however, that the number-one reason for choosing a career in engineering is job satisfaction. Many of my former students (now practicing engineers) have returned to tell me the same story: After the honeymoon period of a new job (i.e., after the initial euphoria associated not only with obtaining an excellent salary and the corresponding increase in one's standard of living, but also with meeting new people, working in new surroundings, etc.), the number-one factor in sustaining you in your employment (and therefore in your career) is job satisfaction—how much you enjoy your job daily. Bright, inquisitive, intelligent people need to be challenged; they cannot spend their working lives in boring, routine, mundane occupations, no matter how high the material rewards. A career in engineering challenges talented people to use all of their skills on a daily basis.

Apart from the rewards, benefits, and opportunities afforded by a career in engineering, you will find that an engineering education will increase your confidence, your general analytical skills, your ability to communicate with people on all levels, and perhaps most of all, your ability to adapt to almost any new situation. You see,

engineering education is a training of the mind!

You can go on and do almost anything after an engineering education. You can become a politician, a movie actor, a teacher, a physician, an entrepreneur, a lawyer, a consulting engineer, or a manager. Look around you. I'll bet you can find at least 10 examples of people who began their careers with an engineering education and ended up being extremely successful in a completely different career.

2.4 EQUIPPING YOURSELF FOR ENGINEERING STUDY

In Chapter 1, we discussed the keys to success in engineering study. These strategies are vital for excellent performance throughout your pursuit of an engineering degree. We will return to them again in subsequent chapters when we discuss *specific* strategies for maximizing performance in engineering courses. In this section, we focus on what you should do to ensure the best possible start to your engineering degree.

Many of you were high school students only a few months ago. As recent graduates, you are equipped with attitudes, work ethics, and strategies that worked well in high school. Now you are freshman engineering students. Does it follow that you can continue to operate as you did in high school? The answer is a resounding "no." There is a huge (social and academic) gap between a high school and a university, and it seems to be growing bigger. This gap exists primarily because of the following reasons:

1. ***Attending a university is a different ballgame.*** A university is not equivalent to the next year of high school. It's a life change. There are new pressures, new people, new competitors, and new standards. All of a sudden you will find yourself surrounded by people as bright as, if not brighter than, you are. Standards previously accepted as excellent become good or average. You find out quickly that high school routines, attitudes, and philosophies don't work anymore. You might be homesick, you might have financial problems (perhaps it's the first time you've left home), and your parents might even be pressuring you to be the best student out of the 500 or so in your freshman class. Everything is different—a whole new world.

2. ***You probably don't have the required skills to be successful in a university.*** Engineering is a demanding curriculum. If you've come straight from high school, then you've never done anything like this before. Consequently, you have to learn a whole new set of skills, skills that you were not taught in high school, mainly because you never had any reason to use them there. For example, you need to develop appropriate:

   Study skills and strategies

   Attitudes

   Communication skills

   Team skills—the ability to work as part of a team

   Time management skills

   These are absolutely essential to manage the hectic, often crazy, pace associated with the freshman year in engineering.

3. ***What professors assume you know is often not what you've been taught in high school prerequisites.*** Universities ask that you satisfy certain prerequisites before they allow you to register for a particular course or program. In doing so, they assume that on Day 1 of the university course, you are ready to take up where you left off at the prerequisite level. The prerequisites are usually high school courses. Sometimes the curricula of high school courses change and no one tells the professors. Thus, a situation arises in which the professor assumes that you know what you don't know, and you're in trouble right from the start.

4. ***What professors assume you can do is often not what you really can do.*** You forget things. The summer between high school and university is three months long. Worse still, you may have had a significant period of absence from school. In either case, your prerequisite skills will not be as fluent as they should be (indeed, as your professor will assume they are), and you probably won't be ready to use them on the first day of classes. I like to compare this situation to that of a long-distance runner who has had three months off and then tries to run a marathon without any training. What are his or her chances of even finishing? Not very high! As with most challenges, you have

to be in shape to take on a new engineering course. You must be fluent in the prerequisites (have them at your fingertips) and be ready to build on existing knowledge.

Most first-year engineering students begin to feel the effect of the gap around midterm time, when grades begin to tumble and expectations are far from realized. Recovering from this point is not easy: It takes a lot of effort to change your study habits and lifestyle amid the usual chaos associated with five or six courses in full-steam-ahead mode. Make sure you don't find yourself in this position by preparing ahead of time. The gap is not necessarily your fault, but the responsibility for filling it rests with you. The good news is that it's not difficult to do so; successful engineering students have been doing it for years. The key is to learn what the best students have done and to do the same yourself. We will discuss this in detail in Chapters 5 and 6. For now, what follows are a few samples of general things that you can do to prepare yourself for engineering study:

1. ***Expect changes and adapt.*** Unfortunately, most freshman engineering students don't find out that they lack the necessary success skills until well into the first semester, when it's too late. You can avoid this scenario by accepting the fact that you will have to learn a whole new set of skills and preparing to change the way you do things before the semester begins. If you expect things to change, you will be ready for change, and that's an advantage. Use the initial few days and weeks of the semester to adapt your study habits and your everyday routine to the demands of the engineering curriculum. After all, the latter should be your first priority—make it so.

2. ***Ask.*** Make it a goal to find out as much as you can about how to be most effective freshman year. Talk to professors, advisors, and any second-year students who have had a successful first year, and find out as much as you can about how to maximize your performance. This is an excellent investment of time and effort.

3. ***Get together with other first-year students.*** Share problems, concerns, and information with your peers. Not only will this help alleviate stress and foster friendships, but it will lay strong foundations for the inevitable teamwork that comes with engineering study.

4. ***Get organized and manage your time effectively.*** In the initial stages of an engineering degree, it is of the utmost importance to get and stay organized. Use a notebook or calendar to write down your appointments, commitments, and obligations. Don't try to remember things: There is just far too much information to remember. Make lists of things to do for short-term (e.g., daily) and long-term (e.g., weekly) goals, and cross off each item on the list as soon as it has been achieved (this part is extremely satisfying). In addition, you must learn to manage your time. Schedule blocks of time for studying and blocks of time for other things (e.g., recreation or family responsibilities). And stick to your schedule! Decide beforehand what you will do during a particular period of time and where you will do it (e.g., in the library, at home, etc.). There is nothing worse than spending half the allocated study block deciding what and where to study, so always do this well in advance. For example, you might reserve the period from 2:00 P.M. to 4:00 P.M. on Thursday afternoon to complete the statics assignment in the library.

Or you might reserve the period from 11:00 A.M. to 11:30 A.M. to see your professor regarding some problems with the calculus lecture. Again, write it down in your calendar or planner. Be meticulous, and soon you will become a master planner, making the most effective use of your time. You will establish routines that will greatly improve your effectiveness, and you'll have much more fun! (More on this in Chapter 5.)

5. ***Before a course begins, know what your instructor assumes you know.*** One of the main reasons qualified students do not do well in engineering is that they approach new courses from the cold-start position, not realizing that most courses simply pick up from the end of the prerequisite course (usually the previous semester). These students spend the first half of the new course relearning the material from the prerequisite course. It has been shown time and time again that a few hours' review of the most important prerequisite skills before the course begins will significantly improve one's performance during the course. Consequently, ask your instructor to identify the most important prerequisite skills an effective start in the (new) course. Make sure those skills are fluent and in working order. You may have forgotten the most important prerequisite techniques (particularly if they were covered last semester), so practice with some warm-up exercises if necessary. For example, in calculus, you might practice some factoring, trigonometric identities, or the algebra of functions; in engineering mechanics, you might recall equations of motion and practice drawing some free-body diagrams. Remember, most engineering subjects are cumulative: One part usually depends heavily on a knowledge of the previous part. Consequently, it pays to ensure that your prerequisite skills actually work. (Your instructor will assume that they do.)

2.5 COOPERATIVE EDUCATION PROGRAMS (CO-OPS) AND INTERNSHIPS

Most engineering schools offer two types of degree programs: a traditional program and a cooperative education program. In the cooperative education program (co-op), students complement their academic studies with relevant and productive paid work experience from employers in business, industry and government. Typically, in a co-op program, you would alternate periods of work with periods of study. In most cases, because of the alternating periods of work, co-op programs are longer than traditional engineering programs lasting perhaps one more year than the traditional program.

Since cooperative education programs are degree programs in their own right, they require separate registration and administration. These are usually handled by the engineering school's own *cooperative education center* which gives assistance to co-op students in all aspects of the co-op program including assistance in finding relevant work-term employment for the duration of the program.

Internships are also career related work opportunities. They are usually sponsored by companies working in a range of engineering disciplines. Internships, however, may be paid or unpaid and are most often one time, rather than alternating, work experiences occurring usually over the summer months. Some companies however, may also offer part-time internships during the fall and spring semesters. Most companies insist that you have completed a minimum number of course credits before considering an internship. Usually this means that you can apply for internships from the summer of your sophomore year until graduation. Unlike co-op programs, internships generally carry no academic course credit and do not require course registration (although there can be exceptions—check with your particular engineering school). Information on internships in your area of speciality can be found in your career services office or in the administrative offices of your engineering school.

The advantages of taking career related work opportunities during your time in engineering study are many and varied. They include:

Advantages of a Cooperative Education Program/Internship

- You gain valuable real work experience which prepares you for the future job market
- You learn on-the-job skills that you cannot learn in the classroom
- You have improved employment opportunities upon graduation
- You get a great foretaste of the work-world allowing you to bridge the gap between academia and the workplace
- You get to apply theory learned in class to real world problems
- You learn to work in groups and interdisciplinary teams
- You gain motivation and enthusiasm for your training as an engineer
- They allow you to make informed decisions when it comes to deciding your career path
- You make valuable contacts for networking inside the engineering and business professions
- You get the opportunity to sample different working environments

Employers also gain from co-ops and internships having access to a year-round supply of well-trained and highly motivated student employees for short-term projects.

Problems

1. What is engineering? Write down your definition of *engineering* and what you think the subject is all about. Give some examples to justify your conclusions.
2. Write a paper (of maybe 500 words) on the engineering discipline that appeals to you most. Include the most important features of the *discipline* that led to your choice.
3. What kind of engineer do you hope to become—for example, an analytical engineer, an experimentalist, a professor of engineering, a test engineer, or one of the other kinds of engineer described in Chapter 2? Why?
4. What strengths and qualities do you think would be required to become a design engineer? An analytical engineer? A research engineer? An engineering manager? A test engineer?
5. Write a one-page paper describing the work of mechanical engineers as opposed to that of civil engineers.
6. What are the most modern engineering disciplines? How did they emerge? Write a one-page paper to answer these questions.
7. Write a 500-word paper on what you believe to be new emerging areas of engineering.
8. Complete the following table by listing five jobs typically associated with each type of engineer:

MECHANICAL	ELECTRICAL	CIVIL	CHEMICAL	INDUSTRIAL	MATERIALS	COMPUTER

9. Which of the specialities within mechanical engineering would be of most interest to you? Why?

10. Repeat Problem #9 for each of the engineering disciplines heading the table in Problem 8.

11. Why did you choose to study engineering? Make a list of 10 reasons used to arrive at your decision.

12. Why do you think that someone with an ability in math and science might choose an engineering education?

13. Write down at least 20 rewards associated with a career in engineering. Rank the top 10 in order of importance to you.

14. Write a 500-word paper on why you want to be an engineer.

15. Get together with a couple of fellow engineering students, and think of a product that would do well in today's economy. Write down what kind of engineering would be required to bring the product to reality and then to the marketplace. For example, what would be the role of the design engineer? The mechanical engineer? The test engineer, etc.? (You do not need to include all the different engineering disciplines.)

16. Which people influenced you the most in your decision to pursue an engineering education? How? Compare your answers with the reasons given in Problem 11.

17. The following questions appeared at the beginning of Chapter 2:

    "What reasons lead you to believe that you are *ready* and *equipped* to study engineering? What are the main differences between studying at a university and studying in high school?

    What new *success skills* do you need to succeed in engineering study?"

    Write down answers to each of the preceding questions.

18. Consider the following first-year engineering courses:

    Chemistry

    Engineering mechanics

    Calculus

    Physics

    vii. Finish this list by adding any other core courses required in first-year engineering by your particular engineering college.

    viii. Look in the calendar for the course descriptions of each of the courses you wrote down in Part (i).

    ix. List the formal prerequisites required to register for each of the courses in Part (i).

    x. Identify what you regard as the most important (*working*) prerequisite *skills* you learned in each of the prerequisite courses listed in Part (iii).

    xi. Ask the professor from each of the courses in Part (i) to identify the most important (*working*) prerequisite skills he or she *requires* right from the beginning of the course.

    xii. How long has it been since you studied each of these prerequisite courses?

    xiii. Do you anticipate having any problems as a result of your answers to Parts (iv)–(vi)?

19. Interview a second-year engineering student. Find out the following:

    i. What new study skills did he or she have to learn that were never taught in high school?

    ii. What did he or she do to manage time effectively in the first year?

    iii. What does he or she think of teamwork or collaborating with fellow students?

    iv. Ask the student to give you the benefit of his or her experience of first-year engineering, and list the most vital skills required to *survive* first-year engineering.

20. The word *calculus* often evokes images of fear and pain in many students. Much of this reaction can be attributed to a lack of *preparation* and a lack of *fluency* in the skills required before entering the course (e.g., students lack or have forgotten skills related to factoring, algebra, functions, trigonometry, etc.). Consider the following questions from *precalculus,* which identify some of the skills I *require* from my students on the first day of a beginning calculus course:

    A. Is $2x^2 + x + 1$ always positive, always negative, or sometimes negative and sometimes positive?

    B. Can you rationalize the denominator of the expression $1/(x^{1/3} - 8)$?

    C. When is the expression $(x - 4)/(x^2 - 8)$ positive, negative, zero, or undefined?

    D. If $f(x) = 3$, what is $f(x + h)$ and $f(x^2)$?

    E. Factor the expression $6(x - 2)^{-1/3}(2x + 1)^{1/3} - 2(x - 2)^{2/3}(2x + 1)^{-2/3}$.

    F. What is the domain of $f(x) = \sqrt{(x^2 - 3x + 2)}$?

    G. If $\cos 2x = 2\cos^2 x - 1$, write down formulas for $\cos 4x$ and $\cos 20x$?

You should be able to answer all of these questions correctly in approximately 15 minutes to consider yourself well prepared for first-year calculus. See if you can answer the questions correctly in that amount of time and then take action to ensure that you have the relevant prerequisite skills (i.e., review the necessary concepts).

21. Talk to your engineering professors and devise assessment tests similar to the one in Problem 20 for each of the first-year engineering subjects listed in Problem 18. The tests should be designed to gauge your level of preparedness for each of the courses; thus they should cover the required working prerequisite skills in an appropriate time.

22. Get together with a group of fellow first-year engineering students and devise an assessment test for high school graduates about to enter first-year engineering at your institution. The test should investigate whether the students have the necessary success skills and academic (working) prerequisites required for first-year engineering. Imagine how valuable such a test would be if it could be taken by high school students *before* arriving at your institution.

23. Write a 750-word paper on what a high school student considering studying engineering in the fall should know about being successful in first-year engineering. You should use the information obtained in Problem 22.

24. Go to your local career services office and find out any information about possible internships in your engineering speciality.

25. Write a one page paper outlining why you might benefit from work experience while you are studying engineering.

3

The Role of the University

Universities and colleges have always played a pivotal role in *training* engineers. In addition to providing the environments and opportunities necessary for *learning*, they continually collaborate with professional engineering organizations and engineering accreditation boards to develop engineering programs that are up to date and compatible with the ever-changing needs of modern society.

Your decision to pursue an engineering education has committed you to spending the next four or five years *learning (or training)* in a *university environment.* Consequently, you must learn how to be most effective as an *engineering student.* The first step in doing so is to understand as much as possible about your *learning environment*, namely, its basic structure and how it works. This will ensure that you derive maximum benefit from each of the resources and facilities available as part of your engineering program.

3.1 MAKING EFFECTIVE USE OF CLASS TIME

In general, three different class formats are used in colleges and universities:

1. Lectures
2. Lectures with student participation (active lectures)
3. Tutorials, seminars, or laboratories (Labs)

OBJECTIVES

In this chapter you will:

- Learn about the role of the college or university in your engineering education.
- Learn how to make effective use of the time you spend in lectures, laboratories or seminars.
- Learn how to get help from your engineering professors.
- Find out why liberal arts courses are an important part of any engineering program.
- Learn how to make effective use of campus resources.

All three formats share common objectives: to communicate relevant information and to facilitate learning. However, they realize these objectives in significantly different ways. The key to being effective in class is to know exactly what to expect from each different type of class and to be aware of the expectations placed on you (i.e., to be aware of your role in the class).

Lectures

Lecturing is the most common of all instructional methods used at universities and colleges. In a conventional lecture, there is very little interaction between the instructor and the audience. The instructor spends most of the time presenting the course material by means of a variety of methods (from overhead projectors and computer animations to conventional chalk and board), while the students listen and take notes.

The following is a list of common complaints made by freshman engineering students experiencing the lecture format for the first time:

"The instructor writes too fast!"

"By the time I've thought about one thing, we've moved onto the next thing!"

"All I do is write; I'm not learning anything."

"The instructor should slow down and work through more examples."

"The class is too big—some people are really noisy and I can't concentrate."

"I find it hard to concentrate after the first 10 minutes."

"The class is really boring."

In fact, most of these complaints are valid as regards the typical, conventional lecture. Unfortunately, although it is the most common of the instructional methods, lecturing is usually also the least effective when it comes to learning. So what can you do to ensure that you get the most out of a lecture? The answer is to know exactly what to *expect* from a lecture and to be aware of your role in the lecture.

WHAT TO EXPECT FROM A LECTURE

You should keep in mind that

the instructor has prepared the lecture with your objectives in mind.

In preparing and delivering a lecture, most instructors will do the following:

A. Consult several different sources of information, extract what is most relevant (for your purposes), and present the extracted information in the most clear, concise, and understandable way possible.

B. Ensure that sufficient material has been covered to allow you to solve all assigned problems and any problems likely to be asked in course examinations.

C. Present explanations and illustrative examples of relevant methods and techniques.

D. Indicate standards required and expected in assignments and course examinations.

E. Suggest relevant, targeted practice material.

F. Inform you of any information pertinent to the course or the examinations.

Consequently, lectures are intended as a primary source of relevant information. Do not expect to learn as you go in a lecture (something you're perhaps used to in high school); rather, expect only the presentation of relevant information. Think of a lecture as a source of raw materials (relevant information) to be processed (learned) by you, in your own time and at your own pace.

YOUR ROLE IN A LECTURE

Your primary role in a lecture is, therefore, to gather as much information as possible. Don't expect to learn too much in the lecture itself: The information often comes too fast, leaving little time for contemplation. Instead, concentrate on effective note taking. (More about this will be explained in Chapter 5.) Try to write down everything the instructor writes (this is usually the most relevant information), and take note of anything he or she says. Do not expect to be entertained. Instead, go to the lecture with the understanding that the instructor is there for your benefit, to present to you information that you need to be successful in the course.

To be most effective in the lecture, make sure you adhere to the following guidelines:

- *Arrive prepared.* This means that you should have read all of the assigned material before coming to class. In high school it was often possible to get by without ever having to do any supplementary reading, relying solely on material presented in class. This approach will not work at a university: The lecture provides only part of the information you need. It is your responsibility to find the remainder of the information by ensuring that you read all supplementary materials (from the textbook or otherwise), as indicated by your instructor.

- *Arrive ready and willing to work.* Choose a seat that is free from distractions. Stay away from students who choose to do anything other than work during the lecture. Some first-year engineering students continue to believe that engineering classes are compulsory (as were classes in high school) and that attendance is mandatory. As a result, those students show up reluctantly and with little or no motivation. Instead of applying themselves, they sleep, read the newspaper, or talk with the person sitting next to them. This is a waste of time, effort, and money. Sit as far away as possible from these students, as they will distract you from your purpose. They have not yet realized that the lecture is for *them,* not the instructor.

- *Reread, rewrite, and review.* Within 24 hours of the lecture (preferably as soon as possible, while the material is still fresh in your mind), reread your notes, rewrite any sections that are unclear, and make notes of the things you do not understand. If necessary, see the instructor for clarification and ask any questions arising from the lecture. This is when the learning takes place, not during the lecture. At the end of the process, you will have a clear and concise set of lecture notes, and you will have learned and understood all of the material presented in the lecture. In addition, when it comes to reviewing for a test, you will have detailed notes that you can use to (re)teach yourself concepts that you may have forgotten. (There is nothing worse than trying to understand *scribbles* taken at a lecture several weeks beforehand.)

- *Ask questions at the right time.* Finally, the end of the lecture is usually not the best time to ask questions. There may be a few spare minutes, but both you and the instructor have other classes. It is best (for both of you) to make a separate appointment for asking and answering questions. The most effective engineering students save all their questions for one weekly appointment and use the information they receive to fine-tune the week's lecture notes. (The majority of the rewriting and understanding is performed as soon as possible after each lecture, as previously described.)

Lectures with Student Participation (Active Lectures)

This is the format most often used in high schools and institutions dealing with precollege material. However, it can also be found at colleges and universities, particularly when the class size is relatively small. Frequently referred to as *active learning*, the format covers the material by using a combination of lectures, discussions and question-and-answer sessions. Depending on the size of the class, students may also be organized into groups of two to five and asked questions addressed either to the group or to an individual. This type of class will have you much more actively involved in your own learning and is usually considerably more effective than the typical lecture. The material is generally presented at a slower pace, with time taken for reflection and discussion. Much of what we have said about lectures applies here, except for the fact that the decrease in the lecture component and the increase in the discussion and participation component of the class now gives you more time to understand the material as it is presented. Consequently, you should be prepared to answer (and ask) questions. For example, in class, the instructor may work through a problem prompting you or your group with remarks and questions such as the following:

> *Where did we go wrong in the previous step?*
> *Write down two suggestions as to how we might overcome this difficulty.*
> *How would we adjust this procedure to deal with a more general case?*
> *Write down five physical, everyday examples that make use of this result.*
> *Why do we need to establish this result?*
> *Complete this example and interpret the results.*
> *Why do the results that follow from the theoretical model differ from those obtained in the lab yesterday? Give five reasons.*

Active learning is all about *doing*. It is an extremely effective and productive way of learning. Even if it isn't practiced so much in the classroom, as an engineer, you will engage in active learning almost daily as you *brainstorm* and discuss problems with fellow students. (See Chapter 4.) Discussion and teamwork are the trademarks of engineering students.

WHAT TO EXPECT FROM AN ACTIVE LECTURE

One can sum up in one word what to expect from an active lecture: participation. In an active lecture, you will be actively involved in learning. Consequently, you cannot hide somewhere in the back row or expect just to sit and let the other students in the class do all the work. You will be called upon to participate, so expect to think, respond, and take notes.

YOUR ROLE IN AN ACTIVE LECTURE

In a conventional lecture, you spend very little time actually *thinking* about the material being presented. In an active lecture, the opposite is true: The instructor encourages you to think about the ideas and concepts under discussion. In this way, learning is facilitated inside the classroom, and you and your instructor become *partners* in the learning process.

The key to maximizing performance in an active lecture lies in *participation*. Even though you feel shy, jump in with both feet. Once you get going, you'll find it difficult to stop! In the event that the class is organized into groups, make sure that you feel comfortable with your particular group. The instructor will help you in this respect. Once you work as part of a team, don't simply rely on other team members. Remember, this type of learning depends on participation, so speak up.

Finally, it is important to prepare well for active lectures. Review the material before the lecture, and anticipate questions ahead of time. Familiarizing yourself beforehand with the topic under discussion will give you the edge in class.

Tutorials, Seminars, or Laboratories (Labs)

Recall the lecture format, in which an expert lectures to a group of students, is least effective when it comes to learning. However, lecturing is relatively easy and cost effective, so it remains the most popular method of instruction.

The most effective way to learn anything is through one-on-one instruction. In an effort to provide as much one-on-one instruction as possible, the modern-day teaching system combines lectures with smaller discussion groups, known as tutorials, seminars, or laboratories. These offer you the opportunity to engage in a limited amount of one-on-one instruction and are therefore extremely valuable. The other main objectives in a tutorial, seminar, or laboratory are:

i. To reinforce lecture material with worked-out examples, practice problems, and discussion or active participation. (Remember, we learn by seeing examples and doing exercises.)

ii. To provide an opportunity for you to ask questions and discuss points made in the lectures.

iii. To provide hands-on experience with the application of engineering to real-world problems.

iv. To conduct experimental work to support theory discussed in class.

v. To discuss supplementary course materials or to provide hands-on experience with software packages and their use in related areas of study and work.

Usually, an instructor will assign a particular set of problems to each tutorial, seminar, or lab. These problems are chosen to reinforce lecture material and afford practice for a forthcoming homework assignment or test. Each problem set is almost always assigned well beforehand. This is done to allow you preparation time so that you can spend the corresponding time in the lab or tutorial asking questions one-on-one with your instructor. Such one-on-one instruction is an extremely valuable component of your engineering course. It is an opportunity to ask as many questions as possible on specific examples. Because it is so important, you must prepare appropriately for it. Attempt each assigned problem, note any difficulties, and make a list of specific questions for your instructor before going to the tutorial, seminar, or lab. The most effective engineering students think through each problem ahead of time in order to spend the majority of tutorial or lab time asking questions. (The problems are designed to be challenging and thought provoking.) The least effective students show up unprepared, usually without having even looked at the assigned material. As a result, they spend most of the lab trying to figure out details that they should have considered on their own time (e.g., at home or in the library). They almost never ask questions because they are too busy doing what they should have done beforehand, and consequently, they waste a valuable opportunity for one-on-one instruction. Never forget that an opportunity to ask course-related questions, one-on-one, with someone who knows what you need to know is an extremely valuable commodity in engineering study, one that should never be wasted. Those of you who have ever hired a lawyer know that time spent in consultation is expensive and should therefore be directed towards discussing only relevant facts;

things that you can do yourself, on your own time, you do before the appointment with the lawyer. This way, you pay less money and maximize your effectiveness. Think of tutorial and lab time accordingly. In the world of academics, one-on-one access to an instructor is extremely valuable, particularly when *you* can choose which questions to ask.

3.2 MAKING EFFECTIVE USE OF THE ENGINEERING PROFESSOR

To make effective use of your engineering professors, you need to understand the role they play in the university system. Firstly, the engineering professor does not spend all of his or her time engaged in teaching-related activities. In fact, the engineering professor's time is distributed among three major areas, all regarded as equally important:

a. ***Teaching.*** The typical engineering professor will engage in both undergraduate and graduate classroom teaching, course development, and supervision of related labs, tutorials, and seminars. In addition, the engineering professor will personally supervise student projects (mainly undergraduate) and theses (mainly graduate).

b. ***Research.*** Most engineering professors engage in scholarly research in their own areas of interest. Related to this are duties associated with publishing papers in scholarly journals, supervising graduate students, presenting results at conferences, and applying for research grants to support current and future research. The last of these is not as trivial as you might think: Grant applications can take months to prepare as they require large amounts of information, appropriate proposals, and corresponding budgetary information. Professors usually apply for several grants each year, making this an extremely time-consuming activity.

c. ***Service.*** Professors are often asked to advise formal government departments, industrial and professional boards, and community organizations. They are also actively involved in the operations of their own particular department or faculty, for example, academic planning, staff selection, and curriculum development.

The professor's extensive experience of, and involvement in, so many areas allows him or her to help you not only with your course work (through one-on-one instruction) but also in the following areas:

Advising   Most engineering professors use their knowledge and experience to advise students on both academic and nonacademic aspects of a university education.

Career Counseling   Engineering professors have many contacts in industry, government, and other educational establishments. Some have industrial experience, while others act as engineering consultants. This puts them in an ideal position to advise you on career opportunities and help you find temporary employment—for example, over the summer.

Professional References   Professors will also help you obtain employment, scholarships, or memberships in prestigious organizations by agreeing to write you letters of reference or recommendation. These usually carry a lot of weight, not only because of the professors expertise, but also because of his or her knowledge of your ability in engineering.

As far as you are personally concerned, probably the most important duty your engineering professor performs, beyond classroom instruction, is giving help, one-on-one, during office hours. In this respect, it is interesting to note that many first-year engineering students will not seek help from professors for fear that they will inconvenience them. What these students fail to recognize is that the professor fully expects and encourages students to visit his or her office to ask questions. Asking questions is an essential part of the learning process and extremely rewarding for both parties. As an engineering professor, I am never more pleased than when I see a satisfied student walk out of my office with a firm grasp of a difficult concept. That said, however, there are certain ways to ensure that the time spent in consultation with your professor is used as effectively as possible.

Firstly, it is important to recognize that, in view of the many duties required of an engineering professor, he or she may not always be available to answer questions. To guarantee time for questions, most professors will assign office hours to a particular course. This is a block of time set aside for the sole purpose of answering your questions. During that time, the professor is happy to discuss any aspect of the lecture material, talk about the assignments, and answer any other questions relating to the course. If, for some reason, you cannot meet with your professor during the assigned office hours, you can ask for a separate appointment at a mutually convenient time. Whichever way you choose to meet, there are certain guidelines you should follow to ensure the most effective use of time spent in consultation. The following are some "dos and don'ts" of getting help from your professor:

DO:

- Arrive on time, either during the assigned office hours (preferably not at the last minute) or at the time of your prearranged appointment.
- Arrive prepared and organized. Know exactly what you need to ask. Prepare a list of questions beforehand (if necessary) and bring any required supporting materials, for example, a textbook or the list of questions.
- If you intend to ask a question about a particular problem, bring your work so far; that is, have a clear and methodical written account of your attempt at the problem, indicating where you got stuck and what method or argument you used to arrive at this point.
- Keep your questions targeted and focused.
- Be professional, polite, and courteous at all times. If you do not understand a particular explanation, ask (politely) to have it repeated.
- Ask or encourage the professor to write down any help or explanation. That way you have a record of your visit, and later you will have an opportunity to review what was said in a more relaxed atmosphere. In this respect, it is always a good idea to have a pen and a blank piece of paper available to present to the instructor when he or she begins to explain something. Engineering is primarily a written language—we remember very little of spoken explanations—so make it easy for the instructor to write down his or her explanations.

DON'T:

- Show up outside of office hours or prearranged appointments. More often than not, if you try to meet with your professor during nonoffice hours without an appointment, you will have a wasted journey. Professors allocate their time carefully according to their various duties.

- Ask unfocused, general questions, such as "How do I do Problem 3 of the assignment?" This gives the impression that you have not thought about the problem and wish the professor to do your work for you! Instead, present your (clear and logical) attempt at the question, and ask your professor's advice on, for example, the next step or where you may have gone wrong. The professor is more likely to provide constructive assistance if you do this.

- Ask for help on a problem that you have not thought about. Again, this is basically asking the professor to solve the problem for you. First try the problem yourself.

- Argue with the professor. Instead, maintain a good working relationship with him or her.

- Be afraid to ask as many questions as you need to ask. Just remember to be polite and professional and follow the points previously made. More often than not, professors will interpret your visit during office hours as being indicative of a conscientious effort to do well in the course. This is usually to your advantage when the instructor comes to assessing your overall performance in the course.

Following these simple rules will go a long way toward ensuring that you get your questions answered quickly and effectively and that you maintain a good working relationship with your instructor.

Finally, don't forget that professors are human beings, and they, too, have feelings. Accordingly, make an effort to be nice. Try to smile, listen attentively, and address the professor with the correct pronunciation of his or her name. If you think the professor did a good job in yesterday's lecture, then say so. Also, don't be afraid to pass out compliments on the course, a particular lecture, or even how the professor's teaching has helped you.

3.3 WHY TAKE LIBERAL ARTS COURSES?

"I came to university to study engineering. Why do I have to take liberal arts courses?"

This is a question asked by many undergraduate engineering students—in particular, first-year engineering students. They ask the question mainly because they are focused on the technical side of science and engineering and often don't understand why and where liberal arts courses belong in an engineering education. They view these courses as risky (believing that their strengths lie only in the technical side of the engineering sciences) and as a distraction from their chosen field of study. This is unfortunate, since after all, the decision to include liberal arts courses in an engineering education is made by experienced engineering educators in collaboration with professional engineering organizations and practicing engineers. In this section, we examine the reasons for taking liberal arts courses and how they will benefit you as part of your engineering education.

Liberal arts courses include a range of courses chosen from the arts, management, law, and languages. There are two main reasons these courses are included in an engineering curriculum.

Engineering Is an Education

When athletes train for a particular event, they don't just train the specific set of muscles required to excel in their chosen speciality; they train the whole body. For example, sprinters don't just sprint; they use weights to train their chest, arms, and legs, they run long distances to improve their aerobic capacity, and they practice good nutrition. Athletes recognize that the development of any specific skill is greatly influenced by overall strength and fitness. This is true also when it comes to intellectual skills. In particular, your performance in engineering will be affected by your overall intellectual strength. Liberal arts courses contribute to that intellectual strength by exposing you to ideas, strategies, and practices in arts, communications (see Chapter 9), engineering, economics, humanities, and management. The more you know, the more experience you obtain in as many areas as possible, and the better will be your appreciation of, and ability to use, engineering. After all, you never know when an idea from what seems to be a totally unrelated area can be used to solve an engineering problem. (How many times have you used logical thinking and problem-solving procedures to solve everyday problems such as balancing your checkbook or dealing with a crisis at home?)

The Big Picture

Practicing engineers often comment that new engineering graduates are too focused on the technical aspects of engineering and lack an appreciation for where engineering fits into the big picture of society as a whole. Part of an engineering education, therefore, is to understand that engineers cannot and do not, work in isolation from the rest of the world. They function as part of a society that is made up primarily of nonengineers, such as economists, sociologists, lawyers, accountants, plumbers, poets, and writers. These same people create most of the demand for engineering solutions, and they are the main consumers of the technology that engineers produce. Therefore, it is essential that engineers be able to communicate their expertise effectively to the many different sectors of society. This, in turn, requires an appreciation and understanding of different attitudes, ideas, procedures, beliefs, and opinions. For example, the design engineer obtains ideas by collaborating with consumers who have little or no technical knowledge. Also, increasingly, engineers must compete in a global marketplace where business relationships depend significantly on the understanding of different cultures.

Liberal arts courses will help impress upon you the fact that engineering is a valuable component of society by allowing you to step out of the engineering domain and view your education in the context of its position in society.

3.4 USING CAMPUS RESOURCES

Over the years, engineering schools have developed their resources in response to the needs of the many successful engineering students who have gone before you. Consequently, campuses across the country offer assistance in every conceivable problem (social and academic) that you are likely to encounter throughout your four or five years of engineering study. In other words, your needs have been anticipated, and school resources have been targeted towards maximizing your effectiveness as an engineering student. All you have to do is familiarize yourself with the resources available on campus and find out how to access them.

The following is a list of some of the campus services available to you:

Learning Resource Center

A learning resource center offers assistance in reading, writing, mathematics, computer skills, and study skills. This is usually done through supplementary, noncredit instruction, either one on one or in small groups. The aim here is to equip you with sufficient skills and information to be most effective in your regular classes. For example, if you're having difficulty taking tests or would like to learn more about time management, the learning resource center will offer assistance through qualified professionals experienced in that area.

Student Counseling

Qualified counselors can help you with all kinds of personal problems, both social and academic. For instance, if you are having financial problems, family problems, or even legal problems, a counselor will be able to advise you on your best course of action. Similarly, if you are experiencing test anxiety, memory lapses, or an inability to concentrate in class, an experienced counselor may be able to refer you to the appropriate health expert. Sometimes, counselors just sit and listen. Often, sitting down and talking about your problems with a friendly and concerned individual is all it takes to get you back on track.

Office of Student Financial Aid

This campus resource offers information and advice on student loans, scholarships, and grants. It may also provide emergency loans to students who find themselves in severe financial circumstances.

Student Health Services

Student health services offer complete health care (both physical and psychological), often under one roof.

Career and Placement Services

Career and placement services provide information related to the search for both summer employment and full-time employment. Advice is available on all aspects of the job experience, from preparing a resume to being effective in job interviews. Look out for workshops on these and related topics. Another extremely attractive resource offered by these campus services is a database of information on companies, industries, and professions.

Campus Libraries

As excellent sources of reference materials as well as great places to study, campus libraries have a unique atmosphere that is highly conducive to learning and free from unwanted noise and distractions. Campus libraries usually conveniently situated (allowing you to study between classes), and they open early and close late, giving you lots of time to plan study sessions.

Remember, campus services will not come to you: It's up to you to seek them out. Invest a little time in reading about these and other services in your campus literature (calendars, student handbooks, etc.) That way you can prepare for any eventuality.

Problems

1. Describe the role of the professor in a lecture.
2. Describe your role in a lecture.

3. Write down five things that you expect from a lecture. For each of these, describe your role in achieving these expectations.

4. For Problem 3, describe the role of the lecturer in achieving your expectations.

5. Write a one-page paper entitled "Who Is Responsible for my Learning?" Address issues such as the relationship between what you learn and what your professors teach: Are professors responsible for teaching you, or are you responsible for teaching *yourself* with the help of professors?

6. Suppose the professor in one of your courses is a terrible lecturer. He is inaudible, unclear, and disorganized, and always makes mistakes in class. How would you deal with this situation? Make a list of all possible actions and the corresponding advantages and disadvantages of each. Formulate a strategy for dealing with the situation. (Bear in mind that the professor has the information you need to be successful in the course.)

7. What skills are required in making effective use of your professors, particularly when you ask questions? Describe how any or all of these skills will add to your abilities as an engineer.

8. How would you deal with an unfriendly, uncooperative professor who seems rude and unwilling to help you? Make a list of strategies, and devise a plan for dealing with such a professor.

9. What would you do if, after three visits to see your professor, you still cannot grasp an extremely important point made in yesterday's lecture? Make a list of strategies, given that your objective is to get the relevant information.

10. Write a 500-word paper on "Why Liberal Arts Courses Are an Important Part of My Engineering Education."

11. Identify each liberal arts course in your curriculum this year, and write down how each will benefit your engineering education.

12. Consider the various campus resources discussed in Section 3.4. Find the corresponding services on your campus. Select the three that you regard as the most important to your engineering education, and visit their offices.

13. Suppose you found yourself in the following situation: You just can't seem to finish any particular test on time. On some occasions, you find yourself only halfway through the test when time is up. You know how to do all the questions, but you just can't seem to get them done in the time allotted for the test. Write a one-page paper on what you would do in this situation.

14. Suppose you find yourself falling asleep daily, midway through the 11 o'clock class. This is beginning to annoy the professor and affect your performance in the course. What do you do?

4

Learning in the University Environment

High schools and universities offer vastly different educational environments in terms of the way in which knowledge is delivered and the amount of independent learning that is expected. In high school, there is very little need to find things out for yourself. Teachers tell you everything you need to know to solve your problems. You learn by absorbing this information and repeating it on homework assignments and examinations.

At a university, your role in the learning process is much more significant. There is no longer one central source of wisdom. Instead, information has to be sought out, often from many different sources. In addition, you may have to deal with the usual imperfections in the system, such as poorly prepared and delivered lectures, confusing and uncooperative instructors, and unreadable textbooks. This means that you have to take the initiative and find out for yourself what you need to know. In other words, you are expected to

Take responsibility for your own learning!

OBJECTIVES

In this chapter you will:

- Learn about the different teaching styles used by different professors.
- Learn how to identify your own learning style and what to do if it doesn't match your professor's teaching style.
- Discover the benefits of teamwork and group study.
- Learn about the benefits of being actively involved in student organizations.
- Find out about engineering ethics and your college or university's code of student behavior.

This is the fundamental assumption on which the postsecondary education system is based, and until you realize it, you cannot maximize your performance in engineering study.

Most engineers are forced to become more *independent* in their learning after they graduate, usually on their first job, when they are faced with a collection of real-world problems, most of which are poorly defined and without any known solution. This time there are no professors to ask and no lectures or textbooks that will reveal the answer, so they learn to find the necessary information themselves. In effect they acquire the skills of thinking independently and learning independently, two of the most important skills an engineer can posses.

Wouldn't it be great if you could learn these skills immediately so that you can put them to use right now? This chapter will help you do just that. It will provide you with training in independent learning ahead of time, at the very beginning of your university education. This will serve not only to prepare you for life after graduation, but also to ensure that you perform at the highest level throughout your engineering education.

4.1 LEARNING AND TEACHING STYLES

I vividly remember my first job after graduation. I was assigned to work as part of a team developing instrumentation for use in military aircraft. The project leader was an ex-air-force commander called Stan. He was an electrical engineer by training, but had extensive experience in aerodynamics and applied mathematics. My job was to develop a mathematical model for the aircraft in flight, so that we could test onboard measuring instruments in the lab. To do this, I needed to know how the various motions of the aircraft would affect the different variables used in my model. I knew more about vectors and geometry than I did about aerodynamics, so I took my mathematics to Stan's office and asked him how I should adjust the equations so that my model adequately represented the aircraft's motion. Stan pulled out a diagram of the aircraft and began to describe how air resistance, sideslip and, aircraft flutter contributed significantly to the overall motion of the aircraft. He waved his hands and arms about a lot and, at one point, picked up a model of an aircraft he had on his desk and began to demonstrate the different motions and how they are influenced by the aircraft's design and orientation. I left the office about an hour later and none the wiser. It was as if Stan and I spoke completely different languages. I enjoyed speaking to him, and his explanations were clear and extremely interesting, but I just couldn't see the connection between what he was *telling me* and what I *needed to know*.

The problem was not one of communication (I had understood what Stan had told me), but rather one of a mismatch between Stan's *teaching style* and my *learning style*. Stan was very much a *visual person*. He preferred to use demonstrations, pictures, and diagrams. He wrote down very little and avoided abstraction whenever he could—he preferred to stay in the real world at all times. I, on the other hand, had spent four years learning in the typical classroom situation, where the external information is presented in the form of written and spoken words. I was accustomed to lectures, equations, and theory written either on the overhead, on the chalkboard, or in my textbook. In other words, I was mainly *verbal* in my learning. The information Stan presented to me was indeed what I needed, but it was in the wrong form for me to use. Immediately after the meeting, I reviewed and expanded the notes I had taken during the previous hour. I then began to try to translate them into my language using various books, reports, and articles and talking to different engineers. It was hard work, but eventually I succeeded. I learned a lot that afternoon. In particular, I learned that if I was to be successful in engineering, it would be up to me to make information fit into my particular mode of learning. I couldn't expect others to change for me. How about you? What is your preferred mode of learning? For example, would you rather see a video demonstrating how the flow of air over an aircraft wing allows an aircraft to fly, or would you prefer to read about Bernouilli's equation and the theory of flight? Do you learn more effectively from *visual* or from *verbal and written* information? It is important to know as much as possible about how you learn most effectively. That way, if there is a mismatch between your *learning style* and your professor's *teaching style* you can take the necessary steps to make up for the shortfall.

Your *learning style* is determined largely by how well you *receive, respond to, and process* different forms of external information. For example, there are a number of different groups of learners [1]:

Visual learners   Visual learners learn more effectively through the use of demonstrations, pictures, graphs, sketches, and similar *visual representations.*

Verbal learners   Verbal learners respond more to the written or spoken word. They like to read about things or hear explanations from an expert.

Sensing learners   Sensing learners focus on things that can be sensed, that is, what is seen, heard, or touched. They like *facts and data, the real world,* and above all, *relevance.* They are patient with details and enjoy solving problems by standard methods. (To a sensing learner, the *answer* is of more interest than the *solution.*)

Intuitive learners   Intuitive learners are dreamers. They prefer ideas, possibilities, theories and abstractions. They look for meanings, prefer variety, and dislike repetition. They miss details, make careless mistakes, and often don't check their work.

Active learners   Active learners tend to process information while doing something active (for example, talking). Consequently, active learners think out loud, try things out, prefer group work, and generally learn by *doing* the thing for themselves.

Reflective learners   Reflective learners think to themselves, prefer working alone, and want to understand or think things through before attempting to do anything for themselves.

None of us fits exclusively into *just one* of the preceding categories. We all share elements of *each of them,* but our preferences differ. For example, Stan had a preference for visual, sensing, active teaching and learning, while I tend more towards the verbal, intuitive, reflective type of learning.

In engineering school, most professors tend to teach:

---

verbally (by means of lectures, overheads, the chalkboard, and textbooks).

intuitively (using words, mathematics, and theory)

passively (nonactively, without student participation).

However, when learning, many engineering students tend to be:

---

visual (learning most from demonstrations, pictures, diagrams, sketches, and graphs).

sensing (learning most from practical, real-world applications, relevant examples, worked-out examples, facts, and data, not just theory).

active (learning most by discussing, thinking out loud, working in groups, collaborating, and experimenting)

A mismatch between *learning and teaching styles* can have drastic effects on your performance in the classroom. For example, you may find yourself becoming increasingly bored, inattentive, or perhaps disruptive to those around you (in particular, those with learning styles that do match the professor's teaching style). As a result, you may do poorly on assignments, and examinations and perhaps begin to lose enthusiasm and the all-important motivation required to be successful in engineering.

So what can you do if you find yourself in a situation where your learning style does not match the professor's teaching style? What follows are a few simple suggestions based on my experiences as both a student and an instructor.

1. **Don't** label the professor as a "bad teacher" and then blame him or her for all your problems. Complaining will do very little to help you. Remember, the professor is not responsible for your learning. *You are!* Instead, take matters into your own hands and use your energies constructively, as follows:

2. **Find out** what you need to make the course material more compatible with your particular learning style. For example, do you need more pictures, more demonstrations, more worked-out examples, more real-world applications, more theory, more formulas, or more corroborating evidence? Find out what you need.

Having problems in a class? Don't panic—get help!

Figure 4.1. Typical college class.

3. ***Talk*** to your professor about the difficulties you are experiencing. (See Section 3.2.) Suggest ways in which the professor could help you get more out of the lectures. For example, you might ask the professor to:

    a. Add more worked-out examples to the lectures.

    b. Illustrate important concepts with real-world applications.

    c. Provide some demonstrations of how a particular theory works in practice.

    d. Suggest any additional resources that might help you process the necessary information more effectively (e.g., books, articles, videos, Web page addresses etc.).

    Remember, your objective here is not to persuade the professor to change his or her teaching style (which would be almost impossible). Rather, it is to ask the professor to provide you with the necessary information to translate the course material into a form more *compatible* with your learning style and to fill in any gaps (as I did with Stan).

4. ***Talk,*** discuss, and collaborate with people who are likely to know what you need to know (e.g., graduate teaching assistants, bright or resourceful classmates, former professors with whom you have enjoyed a good working relationship, etc.). Sometimes an alternative explanation delivered from a different point of view will make things clearer. For example, I would often ask my mathematics professors to explain difficult concepts from engineering mechanics. This worked for me because I was more *intuitive* in my learning and I needed to *see* certain concepts through the eyes of a mathematician. (For instance, to me, relative motion analysis was an exercise in vector geometry with applications to mechanics. To others, relative motion analysis was an exercise in mechanics with applications to vector geometry.)

5. ***Consult*** sources that will *supplement* or provide alternative explanations of information from the lectures. Seek out other references on the same subject matter (e.g., other textbooks, journal articles, videos, CD-ROM, and anything that gives alternative explanations of the points that confuse you). Colleges and universities are information centers that provide knowledge, both new and old. Usually, if you look hard enough, you can find what you need.

Following these simple steps will not only maximize your effectiveness in class, but will also dramatically develop your ability to *learn independently*—in all aspects of your life, not just academics.

4.2 TEAMWORK: COLLABORATIVE LEARNING

Another crucial step in maximizing your learning effectiveness is to take full advantage of the benefits that come from *learning as part of a group or team.* There are three main reasons for doing so:

1. Teamwork provides the opportunity for collaborative learning.
2. Teamwork keeps you motivated.
3. In the real world, almost all engineering is done by teams.

Let's take a closer look at each of these.

Teamwork Provides the Opportunity for Collaborative Learning Before I came to my present position in an engineering department, I worked as a mathematics professor teaching a range of pure and applied mathematics courses, mainly to freshman science and business students. My office hours were always busy with the usual questions on assignments, the lecture material, and forthcoming examinations. Students would arrive one by one, more or less asking the same questions, which were almost always related to the tougher problems on that week's assignment. When I came to my present position, the first engineering course I ever taught was "Engineering Mechanics II: Rigid Body Dynamics", an extremely challenging and demanding course by all accounts. Consequently, fully expecting to be inundated with questions, I doubled my usual office hours, put my teaching assistant on full alert, and prepared for a barrage of questions. The first couple of weeks' office hours were rather quiet. I attributed this to the fact that my students were still settling into the course. The next two weeks were a little busier, but different: Those students who did come to see me came in *groups*. It wasn't like the endless stream of *individuals* to which I had become accustomed in the mathematics department. Instead, I'd get different groups of students, with an *appointed spokesperson* who would ask the questions. The whole group would then listen to my explanations, ask any follow up questions, and finally, leave together. The questions themselves tended to be less trivial and more involved than those I'd gotten as a mathematics professor. Clearly a lot of thinking had been done before the group had decided to come and seek my help. This pattern continued throughout the semester, even during exam time.

It was in this first engineering course that I learned of the importance of *teamwork* in engineering study. By arranging themselves into groups, my students had introduced the (added) advantage of learning together and from each other. In other words, they were independently practicing good, solid principles of *collaborative learning*. Most of the routine questions and problems were being answered within the group, through discussion, the exchange of ideas, and brainstorming (thinking out loud and developing ideas and solutions collaboratively—see Section 4.3). This explained why my office hours were so different. In fact, they were being used much more effectively.

Dr. R.M. Felder, an internationally renowned expert in engineering education, has the following to say about the advantages of collaborative learning, as well as how to make collaborative learning effective [2]:

> When you work alone and get stuck on something, you may be tempted to give up; when you're working in a group, someone usually can find a way over the hurdle so the work can proceed. Group work also exposes you to alternative ways to solve problems that may be more effective or efficient than your way. Moreover, students routinely teach one another in group work—and as any professor will tell you, teaching something is probably the most effective way to learn it.

> A wealth of educational research supports the effectiveness of collaborative learning. Students who consistently work together on problems in study groups and, when permitted, homework groups, get higher grades, retain what they learn longer, enjoy classes more, and gain more self-confidence than students who only work individually and competitively [3]. Industry is well aware of the power of collaborative work: virtually all engineering projects are done by teams.

> However, simply getting together with some friends to go over problems is not enough to get the full benefit of the team approach. Here are some ideas for making *collaborative learning effective*.

> • *Work in groups of three or four.* When you work in pairs, you don't get a sufficient variety of approaches and ideas and there is no good mechanism for

conflict resolution. When you work in groups of five or more, some group members tend to be left out of the *active problem-solving process.*

- *Outline problem solutions by yourself first.* Often the hardest part of a problem is figuring out how to get started. If all problems are tackled together by the entire group, one of the quicker students may initiate every solution. If that one isn't you, you may think you'll know how to begin solving similar problems in the future—on the exam, for instance—but you probably won't. An effective way to work on a set of problems is to outline the solution to every problem yourself, without doing detailed calculations; then work out the complete solutions in the group.

- *Make sure everyone understands every solution.* Students in a group will often go along with a solution without really understanding it. For group work to be fully effective, every group member should be able to explain in detail every solution obtained in a work session. Having the group members (particularly the weaker ones) go through these explanations before ending the session is a good way to make sure that the session has achieved its objectives.

Teamwork Keeps You Motivated   Let's face it, sitting in silence, by yourself, trying to learn material from a book or from notes taken during a lecture is not the most stimulating of pastimes. It's easy to lose interest, become bored, and then to give up altogether. This is not to say that learning by yourself should be avoided; in fact, it is an essential component of an *effective overall learning strategy.* (See Section 4.3.) Variety in your learning, however, is just as important. Teamwork or group work contributes that variety by providing a constant source of fresh ideas, new approaches, and motivation. It's also fun, which means that you are more likely to do more of it.

People are, without doubt, the greatest motivators of other people. As part of a group, you associate with people who have the same goals, aspirations, hopes, and objectives as you. There is a *synergy* that comes from trying to achieve a common purpose. You don't feel isolated, and you have the technical and psychological support of the other group members. This also has tremendous benefits when it comes to dealing with *stress* or *exam anxiety.* These same benefits are enjoyed by any group of people working as a team. For example, in high school, I played for a variety of soccer, rugby and basketball teams. I remember the pain of training and digging deep to find the motivation to get up at 5:00 A.M. to run over sand dunes with a heavy backpack on. Without doubt, the greatest influence that allowed me to persevere and enjoy an extremely successful amateur career was the encouragement and participation of my teammates, who, by their example, would push me to achieve my maximum potential. When you see someone doing what you want or need to do, you tend to be *pulled along* by their energy. If you think about it, the same is true of almost anything that involves sacrifice and discipline: Whether you are trying to lose weight, get fit, or become a marathon runner, group participation is always an excellent motivator. Have you ever noticed how much easier it is to maintain regular exercise when you exercise with a group rather than by yourself? The same principles apply to engineering study.

Teamwork is the Norm in the Real World   In the real world, everything from making movies to the many tremendous feats of engineering is the result of the efforts of dedicated *teams* of experts. Consequently, as a practicing engineer, it is essential that you develop the ability to work as part of a *team*—not only with other engineers, but also with people from different backgrounds. Remember, team members usually span a wide range of expertise. For example, a team assigned to planning and excavating a tunnel will include experts in everything from engineering science to accounting, management, and ecology. There is no choice in this matter: You must learn to communicate and explain your ideas to others, you must develop the ability to discuss and demonstrate your work, and you must be ready to listen to and appreciate the input of other team members.

In industry (and academia), it is also often the case that those who provide the money to turn ideas into reality are nonexperts who neither understand nor wish to understand the technical engineering aspects of an idea or proposal. In such a situation, it will be up to you to convince potential financial backers that your ideas are worthy of investment. Consequently, you must learn to translate your ideas so that they can be appreciated and understood by nonexperts. For example, try explaining the concept of a *derivative* (from calculus) to a bright 10-year-old. Or explain how an aircraft achieves a takeoff to a poet—do you think you can do this?

Becoming involved in group work as a freshman engineer is the first step in training you for the world of work, and you get to derive all the benefits while you train.

4.3 GROUP STUDY

We learn in high school that we should work individually and competitively and that we're not supposed to share our work with anyone else, since that would put us at a disadvantage. In most beginning engineering courses we are told to complete assignments, projects, term papers, and examinations on our own. It is little wonder, then, that the large majority of freshman engineering students employ learning strategies that promote solitary study rather than *group study.* This is unfortunate, since, as previously discussed, people who study on their own lose out on the benefits of working as part of a group and, hence, some of the major benefits of a quality engineering education.

There is, however, a way in which you can maintain competitiveness, complete assignments, projects, term papers, and examinations on your own, and yet reap all the benefits of working as part of a group. There are two main components involved in doing this:

Information Gathering

This is where you do what's necessary to get all the information you need. In other words, you first collect together the *pieces of the puzzle,* using all the means at your disposal, including:

Attending lectures and labs

Making up for any mismatches in teaching and learning styles

Consulting textbooks and references as required

Working as part of a group (as discussed in Section 4.2)

Completing all assignments

Making effective use of your professors and other sources of help

Brainstorming with fellow students

During this stage, you should discuss, formulate, and collaborate with fellow students on any aspect of your work. Feel free to share thoughts and ideas, and make sure you take full advantage of those offered by other group members or professors.

The last entry on the foregoing list (brainstorming) may be unfamiliar to some of you, so we'll take a moment here to discuss this topic in more detail.

Brainstorming with Fellow Students

Brainstorming is a process by which a group of students generate ideas, suggestions, and solutions to problems of common interest out loud in the group environment. Here are some suggestions on how to conduct a brainstorming session:

- Organize a group of four to six people. Try to include a mix of people with different perspectives and learning styles. (See Section 4.1.) For example, include active learners, intuitive learners, people whose strengths lie in analysis, and people whose strengths lie in interpreting the real world.
- Identify the topic or problem to be solved or discussed.
- Generate as many ideas as possible. Each member of the group should contribute at least one idea.
- Record all ideas, preferably on something everyone can view easily at the same time, such as a blackboard or flip chart.
- Continue generating ideas until the team has exhausted all possibilities.
- Discuss and clarify each idea on the list until you identify the most promising idea or strategy.

To make it easier to conduct the brainstorming session, set a time limit, accept and record all ideas, and allow people to "miss a turn" if they can't think of anything at the time. (This keeps the momentum going.) It is also a good idea not to comment on any ideas as they are being generated—that can be done later, at the discussion stage. Remember also that the most effective group environment for brainstorming is one in which there is a relaxed atmosphere, where it feels more like fun rather than work. In this respect, you may want to have some warm-up exercises to encourage the more introverted group members to break out of their shells and also to set the tone of the session as one of fun and spontaneity rather than serious academic discussion.

Putting Things Together

This stage is the individual stage, where you personalize your contributions and your understanding of the material. Once you have gathered all the required information, techniques, and ideas, sit down by yourself and put together your own contribution. This is where you assemble the pieces of the puzzle. Proceed as follows:

- Organize and collate all your information.
- Understand and explain the information to yourself (as if you were teaching yourself).
- Take detailed and comprehensive notes. (You may need them for review later on in the semester.)
- Produce a finished piece of work (e.g., an assignment, project, term-paper, etc.) in your own words.

Following these instructions will allow you to reap the benefits of group study in all aspects of your engineering education while maintaining your competitive edge and meeting course requirements. For example:

- You can tackle assignments, term papers, and projects as a group, yet submit an individual effort that you understand fully and can call your own.
- You can review for examinations (review material, work through old tests, etc.) as a group, yet equip yourself for success on the exam. (See Chapter 6.)

In other words, you should not feel guilty about group study, and you should not think of it as cheating or as anything other than another tool to help you maximize your performance as an engineering student. Unless you learn to take advantage of all resources available to you, you will never achieve your full potential in engineering. My advice to you is to get out there and join or even organize your own study group, after which you can begin to reap the benefits thereof.

The learning strategies discussed lay out general principles that can be applied to many different aspects of your life, such as, buying a house, buying a car, and looking for a job. These are all significant events that involve finding and processing information, communicating your desires, and collaborating with others who may be able to help you make the best decision. In Chapter 5, we will discuss more specific learning strategies, those designed to maximize performance in actual engineering courses.

4.4 GETTING INVOLVED IN STUDENT ORGANIZATIONS

"Getting involved in different student organizations was one of the best decisions I made as a first-year engineering student. Not only did I make lots of new friends, I discovered my main strengths and weaknesses and learned more about who I was."
(Former engineering student, now a project manager with a large multinational oil company)

Student organizations come in many different shapes and sizes. For example, on a typical university campus, you might find:

- Sports clubs
- Religious organizations
- Social fraternities and sororities
- Academic organizations
- Political clubs and associations
- Culture clubs and associations
- Professional student organizations

As an engineering student, the student organizations that are most accessible and, probably, of most interest to you are the engineering student organizations which operate within your own engineering school. For example:

- Engineering Students' Society. Engineering students' societies are non-profit, student-run organizations which exist to serve the needs of the undergraduate engineering student body. They address the academic, social and professional needs of engineering students. Usually, an engineering students' society is made up of an *Engineering Student Council* and several smaller member organizations from different engineering disciplines. For example, these might include:
  - Mechanical Engineering Club
  - Computer Engineering Club
  - First Year Engineering Students' Association
  - Electrical and Electronic Engineering Club

- Chemical Engineering Students' Club
- Civil and Environmental Students' organization

It is very likely that within your own engineering discipline, you will find at least one member organization representing your interests. The Engineering Student Council is made up of representatives from each of its member organizations and coordinates activities undertaken by several or all of them.

- Student Chapters of Professional Engineering Organizations. Student chapters of professional engineering organizations represent and promote the interests of students within a particular professional engineering discipline. For example, depending on your engineering major, you may choose to join or become involved in the local student chapter of any of the following professional organizations:
  - American Society of Heating, Refrigerating and Air Conditioning (ASHRAE)
  - American Society of Mechanical Engineers (ASME)
  - Institute of Electrical and Electronic Engineers (IEEE)
  - Society of Automotive Engineers (SAE)
  - Institute of Transportation Engineers (ITE)
  - Society of Petroleum Engineers (SPE)
  - American Society of Civil Engineers (ASCE)
  - Structural Engineers Association (SEA)
  - Society of Manufacturing Engineers (SME)

  Many engineering students join more than one organization. For example, as a civil engineering major, you may wish to be involved in the SEA as well as the larger organization ASCE.

- Ethnic and Gender Based Engineering Organizations. These societies have one thing in common: they help provide opportunities for their members to pursue academic studies, professional training and careers within the engineering profession. In doing so, they seek to increase the participation and representation of their members within the engineering profession. It is worth noting that membership in any of these organizations is not restricted to the particular ethnic or gender-based group but rather to those who share the same beliefs and aims of the organizations. Examples of these organizations include:
  - National Society of Black Engineers (NSBE)
  - Society of Woman Engineers (SWE)
  - American Indian Science and Engineering Society (AISES)
  - Mexican-American Engineering Society (MAES)
  - Society of Hispanic Professional Engineers (SHPE)

Why You Should Get Involved in Student Organizations

Before we examine the (many) benefits of joining and becoming actively involved in student organizations, we pause briefly for a reality check. Never forget that:

Your studies are your first priority!

The opportunities to participate in student organizations are numerous to say the least. In addition, the demands (in terms of time and effort) of any particular membership will vary between organizations. Consequently, be careful not to over-commit yourself. Choose your memberships carefully and always remember that your studies come first.

Having said that, let's look at what you will gain by getting involved in student organizations:

- *You Make New Friends*

  You cannot help but meet new people when you participate in a group which shares a common purpose.

- *You Achieve Balance in Your Life*

  Let's face it—no one studies all the time! Participation in a student organization allows you to develop yourself socially and intellectually while contributing to the university community and to the community at large—and, it's fun!

- *You Become More Marketable to Employers*

  Employers are interested not only in your academic ability but also in your potential to demonstrate leadership skills, organizational skills and your ability to adapt to different situations. Student organizations provide opportunities to gain experience in all of these skills. This will give you the edge when interviewing for engineering positions. In addition, as an active member of the local student chapter of a professional engineering organization, you demonstrate your interest in and commitment to your chosen field as well as your willingness to stay up-to-date with recent technological advances in your area. Also, don't forget that the many professional engineers you meet at meetings of local student chapters not only enrich your understanding and appreciation of professional engineering but also make available valuable contacts and opportunities for future employment as a professional engineer.

- *You Discover Your Strengths and Weaknesses*

  The best way to find out what you do best and what you do worst is through exposure and experience. A student club provides relatively safe confines for you to try things out and work on your weaknesses. For example, I have often seen the quiet and shy members of study groups blossom into the most vocal group leaders because of their participation in different clubs and associations.

- *You Discover and Improve Special Interests and Abilities*

  You may have always wanted to try something like boxing, skiing, public speaking or teaching but the opportunity to do so just never came up. It is likely there will be a group or organization in your university which will allow you to do so. Take that chance and add to the educational experience.

- *You Get the Chance to Serve on Committees, Student Government and Contribute to Student Life*

  You can have a real say in the running of your particular student organization or of your university in general by serving on one or more of the many committees/ governing bodies operating in the university environment. For example, departmental committees and faculty committees include student representation and deal with all matters ranging from academic planning to university policy. Student government allows you the opportunity to run for one of the many elected offices providing an excellent opportunity for experience and personal development. If you feel strongly about something, take the opportunity to make your views known—and perhaps change things for the better.

In addition, involvement in engineering student organizations will help you become a better engineering student by improving your study skills. For example, you will

- Develop your ability to work effectively in groups: study groups, project groups and any other group which requires collaboration.
- Develop your abilities in problem solving
- Learn how to communicate effectively with different people—often people with a different background from your own (this skill is greatly valued by employers who will expect you to be able to share your ideas with engineers in different fields and non-engineers and alike)
- Learn how to set and achieve appropriate goals
- Learn how to manage your time, plan ahead and organize your resources in the most effective way
- Form collaborative learning networks—in student organizations, you meet not only other engineering students, often in the same engineering major, but also mathematicians, physicists and a range of other students who might be able to help you in many different ways while you are engaged in engineering study. For example, a group of my own students enlisted the help of a medical student from the same ski club to help them design a wheelchair as part of the senior design project.

As first year engineering students, most of you will not take very many "engineering classes" in your first year. You will be mainly concerned with laying down strong foundations in arts and science for the engineering courses to come in subsequent years. For this reason alone, getting involved in engineering student organization is a great way to maintain contact with engineering and the main reason why you chose to enter university in the first place.

4.5 ENGINEERING ETHICS AND CODE OF STUDENT BEHAVIOR

There's an old Roman saying that loosely translates to.

> Success itself is less important than the path taken to success.

In other words, *how* you achieve your goal is more important than achieving the goal itself. Why should this be? Let's think about it for a minute. Say you had the choice of being treated by one of two physicians. You find out that both physicians graduated from the same (excellent) medical school two years ago in the top five percent of their class. However, after some investigations, you learn that one of these physicians routinely copied work from his classmates and cheated on exams. Would you feel comfortable having that physician care for you while you lay anesthetized in the hospital? Would you feel comfortable with such a person making judgment calls involving your life? Most people don't need to think long to answer that question.

The point is that engineering, as is medicine, is a profession that carries responsibilities. During your time as an undergraduate engineering student, everything you do to achieve your goals will contribute to your training as an engineer, the good things and the bad things. Unfortunately, like most personality traits, any bad habits that you pick up during your time in engineering study will persist after college, when you are empowered with the credentials of the profession.

As part of your undergraduate engineering degree, you will take a course in engineering ethics, which is the study of the moral conduct of individuals involved in engineering. In other words, engineering ethics is concerned with what is morally right, what is morally wrong, and how engineers should behave in situations when it is not easy to decide what is right and what is wrong.

The following are examples of situations in which you may find yourself facing an ethical dilemma:

- You've worked hard on your assignment. One question remains. You struggle for hours, but can't seem to solve it. The next day, one of your fellow students offers her solution for you to copy.

- In an open book examination, you notice that you inadvertently left a copy of the solutions to a practice test in the textbook. Two of the questions appearing in the practice test also appear in the present examination.

- Your best friend tells you that he has been sick the last few days and hasn't had time to finish the assignment and asks if would you let him copy yours.

- During an examination, the person next to you drops his eraser and, while retrieving it, asks you to tell him which method to use for Problem 5.

- Your professor has just returned your midterm examination. He then asks you to fill out the mandatory teaching evaluation form, your opinions on his performance as a teacher.

- You notice one student changing his (already graded) solution to an examination question as the professor reviews the examination in class. This student then asks for and gets five extra points from the professor, claiming that the professor omitted to grade that part of his solution. This now puts you in second place in the class behind that same student.

- In a recent quiz, you discover that the professor incorrectly added the total number of points. You should have scored 11% less than indicated.

- The professor supervising the examination has gone out of the room. Information that you badly need is only a short distance away in your bag on the floor.

- Your girlfriend is also a classmate. She asks you to let her copy your assignments.

- You don't have enough money for this week's rent. You find a wallet containing $100.

In each of these situations, it is difficult to decide what is the right thing to do. What would you do? You have to balance what is morally or ethically right against the immediate short-term benefits to you. For example, if you cheat on an examination, you may obtain a high score on the exam itself, but how accurately does this score reflect your ability or what you have learned? How will you feel about the fact that you obtained your score fraudulently. You may be able to deceive others, but for how long can you deceive yourself?

In first-year engineering, you are concerned mainly with survival, mastering study skills and extremely difficult and demanding material in order to make it into second-year engineering. This is what makes you vulnerable when it comes to ethical issues—ruthlessness in surviving, so often commonplace in college, is often responsible for discouraging ethical behavior. Whenever you are faced with an ethical dilemma of the type just described, try to imagine how you will feel inside some time after you take a particular course of action. Remind yourself that an engineering education is not just about results and grades, but also about developing character and integrity.

There are however, certain actions in academia about which there is no confusion over what is right and what is wrong. These are listed in the university's Code of Student Behavior which can usually be found in the university calendar. This code will list specific forbidden actions that, if undertaken, will lead to disciplinary action, including expulsion, suspension, or academic probation. Among these forbidden actions are the following:

- *Plagiarism*—intentionally claiming other people's works or ideas as your own
- *Cheating*—
  - Obtaining or attempting to obtain information from another student or unauthorized source, or giving or attempting to give information to another student, in the course of an examination
  - Representing or attempting to represent yourself as another person, or having or attempting to have yourself represented by another person in the taking of an examination, preparation of a paper, or other similar activity.
  - Changing a solution or answer on a paper or examination after it has been graded.
- *Confidential materials*—using or distributing any confidential academic material, such as upcoming examinations or laboratory results, without prior consent of the instructor.
- *Fabrication*—falsification of information in any academic exercise (paper, assignment, lab report, etc.).

These are only a few of the academic offenses listed in a standard university student code of conduct. Many more actions (both academic and nonacademic) are regarded as illegal in your university community. You should read your own university's code of student conduct carefully and at least make yourself aware of your responsibilities as an engineering student, trainee engineer, and member of the university community.

Problems

 Consider the following table:

LEARNING STYLE	YOUR PREFERENCE
Visual	
Verbal	
Active	
Reflective	
Intuitive	
Sensing	

Using the information in Section 4.1, mark the styles that most adequately describe how you learn.

2. Using the results from Problem 1, identify your preferred learning style.

3. Consider the following table:

TEACHING STYLE	PROF. 1	PROF. 2	PROF. 3	PROF. 4	PROF. 5	PROF. 6
Visual						
Verbal						
Active						
Reflective						
Intuitive						
Sensing						

Using the information in Section 4.1, mark the teaching styles which most adequately describe each of your engineering professors' teaching styles.

4. Using the information in Problem 3, identify those courses in which your learning style does not match the professor's teaching style. List the consequences of each of these mismatches and what you have done to try to make up for them.

5. Make 10 suggestions on how you think your professors could help remove or substantially decrease the mismatch between their teaching styles and your learning style.

6. Consider the following scenario: In your 8 o'clock class, the professor speaks with a strong Scottish accent and tends to mumble to himself. He often speaks with his back to the class and hardly ever makes eye contact. As a result, you find it very difficult to understand anything he says. In addition, he seems to do nothing more than cover the board with equations, without any explanation of where things are coming from. To make matters worse, the prescribed textbook is awful—very difficult to read and with hardly any worked-out examples! On the last midterm examination, the questions were completely alien, like nothing you saw in class or in the assignments. You scored 35% on that test, which is worth 30% of your total grade in the course! You've been to see the student advisor and the chairman of the relevant department. There is nothing that they can do. You're on your own.

   Write a two-page paper describing the steps you would take to survive this course and maximize your performance on the remainder of the assignments and course examinations.

7. List five situations in which you have benefited from working as part of a group or a team.

8. List five situations in which your performance has increased because of your involvement in group work or teamwork.

9. Have you had any experiences in your life in which you reached a goal because of the intervention of other members of a group? If so, list them. If not, list situations in which you would anticipate that the input of other team members would help you achieve a particular goal.

10. If you do not already belong to a study group or work group, identify two or three people that you think would make good team members. List reasons for your choices. In other words, list the qualities your team members should possess. If you do belong to a study group or work group, list the qualities of the other group members that make the group work well together.

11. Make a list of the different things a study group can help you with.

12. Write a one-page paper about collaborative learning and it's benefits. Title the paper *"The Benefits of Collaborative Learning"*.

13. Solve the following two problems[1] by *brainstorming* and collaborating within your group:

---

[1] Reprinted with permission from National Effective Teaching Institute (NETI), June 1995.

Problem 1 You have two containers, one of which will contain nine liters of liquid when filled to a mark and the other of which will contain four liters when filled to a similar mark. You have unlimited amounts of water available, but no other equipment of any kind. Devise a scheme by which you can measure out exactly six liters of water in the nine-liter container.

Problem 2 An elderly Buddhist monk lives in a small hut at the bottom of a mountain. On top of the mountain is a beautiful temple. The path leading up to the temple is very narrow, just wide enough for a person to walk without obstruction. One morning, just at sunrise, the monk leaves his hut to climb to the temple. He walks at various speeds, pausing occasionally to eat nuts or berries or to meditate. He reaches the temple at sunset. After spending several days in the temple, the monk leaves to return to his hut. He leaves just at sunrise and descends at various speeds, again stopping periodically along the way. He arrives home just before sunset. Prove that there is a particular spot along the path which the monk occupies at precisely the same time of day when he climbs up the mountain and when he descends. (Your proof need not be mathematical.)

14. Recall a recent group study session, for example, one in which you discussed an upcoming assignment. List in the following table (add more rows if you like) which aspects of the session worked well and which were not very productive. Using this information, identify which aspects of your learning strategy would benefit from group involvement and which are best carried out on your own.

WORKED WELL	DIDN'T WORK SO WELL

15. List five different activities in which you engage when preparing for an examination. Which of these five would benefit from group involvement? Why?

16. Write a one-page paper explaining how group study can help you improve your *individual* performance in assignments and examinations.

17. Obtain a list of all student organizations operating at your university. Make a list of all engineering student organizations operating within your particular engineering school. Identify perhaps two or three of these organizations which you have already joined or are thinking of joining. Write down reasons for joining each of the organizations you have identified.

18. Find out as much information as possible on local student chapters of professional engineering societies in your area of interest. Apply to join one of these.

19. Attend a meeting of the local student chapter of the professional engineering organization of your choice. Make a commitment to introduce yourself to at least one other member. Find out what he or she does in the engineering profession. See if there are any opportunities to visit the companies of the professional engineers you meet.

20. Write a one-page paper entitled "The Benefits of Joining Engineering Student Organizations".

21. Suppose you worked on a particular assignment with two fellow students. Between the three of you, you worked through and solved all of the assigned problems. Afterwards, each of you wrote up the solutions, in your own words, ensuring that you understood all the steps figured out previously (within the group). The next day you are approached by a friend who asks you to lend her your solutions so that she may copy them from you. (She was at a party the night

before and didn't get around to doing the assignment.) You refuse on the grounds that it would do her no good to copy the solutions without understanding how to solve the problems. She then says that that's exactly what you did when you worked out the assignment with the other group members. How do you respond?

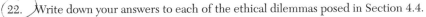

22. Write down your answers to each of the ethical dilemmas posed in Section 4.4.

23. Write a one-page paper describing a situation in which you encountered an ethical dilemma. Mention how you were able to resolve the dilemma.

5

Key Strategies for Maximizing Performance in Engineering Courses

Over the last 10 years, I have taught many different engineering courses at various levels, from an introductory level to a more advanced graduate level. At the end of each course, I approach the most successful students and ask them to describe their study or work habits and any *special techniques* they may have used to achieve that particular level of success. Year after year, the same answers keep coming back. These answers indicate clearly that the most successful engineering students practice, in common, a set of *key study strategies specific to engineering courses*. Perhaps even more significant is the fact that almost every one of these strategies is absent from the study habits of the less successful students. This is no surprise: These same skills used to be taught as part of any basic high school curriculum. Recently, however, weaknesses in the secondary education system have meant that most freshman engineering students arrive without these skills, and very few of them take the time to acquire them for themselves. As a result, many new engineering students find it difficult to make the transition between high school and college. This almost always leads to poor performance in the first year.

OBJECTIVES

In this chapter you will learn how to:

- Organize and manage your time more effectively.
- Prepare for an engineering course (making sure that you know what your professor assumes you know).
- Take notes effectively in class.
- Use the course textbook.
- Maximize your performance on assignments and homework.
- Find and use tutors.

In this chapter, we address the issue of performance and present that very collection of study strategies used by the most successful engineering students. The strategies themselves are sufficiently general to be applicable to *all* engineering courses at *any* level. In particular, we discuss:

- Time management strategies.
- Preparing for an engineering course and making sure that your prerequisite works.
- Effective note taking.
- Making effective use of the course textbook.
- How to be effective on assignments.
- Using posted solutions to assignments.
- Using tutors and study guides.

5.1 TIME MANAGEMENT STRATEGIES

The single most significant change between high school and college is the acquisition of

Freedom!

Suddenly, there is:

- No one to tell you what to do.
- No one to make you do things.
- No one to remind you to do things.

In other words, you acquire a greater measure of *control* over your life. In particular, you gain a greater measure of control over how you *spend your time.*

Time is perhaps our most precious commodity: We cannot create more of it, and we cannot save it up and use it later. All we can do is learn to use the time we have as effectively as possible. Unfortunately, most of us have never been taught how to manage our time. Instead, we tend to learn these skills naturally, later in life, when faced with the usual commitments and responsibilities that come with having a full-time job, a new home, and raising a family.

The need for time management skills in professional engineering is well understood. Only recently, however, is time management being recognized as one of the most important skills required for success in engineering study—despite the fact that the very best engineering students have been practicing it for years! Fortunately, a simple, easy-to-follow time management system is not difficult to establish, and it's one of the most effective ways to give yourself even more time!

Prioritize

Think of your time as you would money. Since you have only a finite amount of money, you have to plan, monitor, and review how you spend it. Most of us do this by formulating a budget: We look at how much money we have and then prioritize our expenditures. First we account for necessities (rent, food, clothing, etc.). Then we distribute the remainder of our money according to our particular needs and circumstances. The same strategy can be applied to managing our time.

As an engineering student, you have three main *priorities* when it comes to spending time.

PRIORITY #1:
COMMITMENTS

These are the necessities, periods of time already allocated as a result of your occupation (in your case, a college or university student). There is no flexibility here; these are things you must do at designated times. For example, you have to attend classes, seminars, labs, and meetings, go to work (part time or otherwise), have meals, and so on.

PRIORITY #2:
STUDY TIME

Study time is required to complete assignments, projects, and papers; rewrite, expand, and improve class notes; read the class text; reinforce the material by performing examples and by finding other sources of information; review for tests, and so on. Study time is *your responsibility* and so is allocated by *you*—not by someone else. Consequently, you have much more flexibility in choosing your study time.

PRIORITY #3:
ENTERTAINMENT,
LEISURE, RECREATION

You should fit your leisure, entertainment, and recreational activities around your commitments and study time. For example, I jog daily, sometimes at 6:00 A.M., sometimes at noon, and still other times in the evenings—basically, whenever I can! Similarly, I'll socialize with friends sometimes at breakfast, sometimes at dinner, or sometimes for half an hour over coffee. I don't refrain from these things—they are too important to me. However, I do give them low priority when it comes to allocating time.

Make a Schedule

Effective time management begins with detailed scheduling. Get yourself a good diary, planner, or calendar, one that will allow you to record your commitments (both long and short term) and schedule your time on a daily basis. There are many advantages to doing this, including the following:

- You have a *visual representation* of your year, month, week, or day, so that you know what's ahead and can prepare accordingly.
- You can *see* if you are using your time as effectively as possible.
- A *picture* of how you are using your time makes it easier to reschedule events or change priorities.
- You are less likely to forget appointments or important events.
- You can note things and forget about them until the appointed time nears.
- You can set up *flags* to remind you of forthcoming important events (e.g., "Exam in one week!").
- You feel more organized and your confidence increases.
- You will receive excellent training for the work world.

Plan from the long term to the short term, as follows:

Long Term

In your diary or planner, mark any important dates for the current academic year. These will be special one-time dates occurring perhaps once or twice a year. For example, you might note midterm and final exam dates, the first and last day of classes, birthdays, and dental appointments, and you might set flags to remind you that certain events are coming up (e.g., two weeks before a final exam is scheduled, you might write "Get sample final exams for review!"). Having these recorded in your planner means that you can forget about them until a later time and be safe in the knowledge that your diary will serve as your *reminder*. I have used this method throughout my professional life as a student and as a professor, and I have found that it is an excellent way to reduce stress and avoid anxiety.

Short Term

It's always useful to be able to take a look at your week or day. That way you can easily reschedule events—for example, to go out with a surprise guest or to attend an upcoming sporting event. Scheduling your entire week also ensures that you don't forget to do anything you are required or intend to do that week. The best way to do this is to use a weekly planner. I have included an example (from a typical first-year engineering student) in Figure 5.1.

WEEKLY PLANNER   The student whose weekly planner is shown in Figure 5.1 is taking six-courses labeled (1), (2), . . .(6). Courses 1, 2, 4 and 5 each have classes on Monday, Wednesday, and Friday, while Courses 3 and 6 have classes only on Tuesday and Friday. There are also various laboratories and seminars associated with different courses. You should schedule your week in the same way, as follows:

- First schedule all your commitments (in bold in the figure).
  a. Classes
  b. Laboratories (referred to as labs in the figure)
  c. Seminars
  d. Jobs (referred to as work in the figure)
  e. Meals
- Next, schedule blocks of time for study. Make sure you use *all available time*. For example, if one class ends at 1:50 P.M. and the next doesn't begin until 3:00 P.M., schedule approximately an hour for study. (Remember, you need to take into account the time it takes to move between classes and the place you choose to study.) You'll be amazed at how productive you can be even in short periods of time, time that might otherwise be wasted in deciding whether or not to study. To do this efficiently, you need to do two things *beforehand*:
  a. Write down *where* you will study. Decide beforehand where the best place to study will be, given the particular circumstances at that time. This will avoid wasting valuable time. For example, if your class finishes at 10:00 A.M., and your next class is at 11:00 A.M., choose a location near either the classroom you just vacated or the one you are about to enter.
  b. Write down *what* to study. Again, so many students waste valuable time procrastinating about what to do during study time. Choose small jobs for short periods of time and longer assignments for extended periods of time. Deciding this beforehand allows you to spend the maximum amount of time actually studying—just get there and get to it!

Making these decisions in advance will improve your effectiveness tremendously. You'll be amazed at how much you can get done during time that would otherwise have been wasted.

- Finally, in the time that's left, try to fit in entertainment, leisure, and recreational activities. You may find yourself working out at different times on different days. That may not make for effective bodybuilding, but that's okay,

	Monday	*Tuesday*	*Wednesday*	*Thursday*	*Friday*	*Saturday*	*Sunday*
0700	B/fast	Workout	B/fast	Workout	B/fast		
0800	**Class (1)**	B/fast	**Class (1)**	B/fast	**Class (1)**		
0900	**Class (2)**	**Class (3)**	**Class (2)**	**Class (3)**	**Class (2)**		
1000	*Research Library/ See Prof*	<u>Class (3)</u> *See Prof*	*Study - Library (Assn #2)*	<u>Class (3)</u> *Study-Lib. (Assn #3)*	*Research Library/ See Prof*		
1100	**Class (4)**	*Study (Library) (Assn #3)*	**Class (4)**	*Study (Library) (Assn #3)*	**Class (4)**	Workout	
1200	Lunch	Lunch	Lunch	Lunch	Lunch	Lunch	
1300	**Class (5)**	**Class (6)**	**Class (5)**	**Class (6)**	**Class (5)**	*See Prof*	
1400	**Seminar (4)**	<u>Class (6)</u> *Prepare - studygrp*	**Lab (1)**	<u>Class (6)</u> *Prepare - studygrp*	*Study- (Assn #5) Room 5*	*Study- Paper #1 Library*	*Study- (Assn 6) Union*
1500	*Studygrp (Assn #1) Room4-1*	*Studygrp (Assn #4) Room4-1*	**Lab (1)**	**Lab (3)**	**Seminar (5)**	Coffee (Mike) Cafe	*Study- (Assn 6) Union*
1600	*Studygrp (Assn #1) Room4-1*	*Studygrp (Assn #4) Room4-1*	Workout	**Lab (3)**	*Studygrp (Lab 3) Room4-1*	*Study- Library Paper #1*	*Study- (Assn 6) Union*
1700	Review Lecture Notes	Review Lecture Notes	Review Lecture Notes	Dinner	*Studygrp (Lab 3) Room4-1*	Dinner	*Study- (Assn 6) Union*
1800	Dinner	Dinner	Dinner	Work	Dinner	Work	Dinner
1900	Relax	Relax	Relax	Work	Relax	Work	*Review Week's Notes -Home*
2000	*Study- Course 1 Home*	*Study- Course 6 Home*	*Study- Course 4 Home*	Work	Fun!	Work	*Review Week's Notes -Home*
2100	↕	↕	↕	Work	Fun!	Work	↕
2200	↕	↕	↕	Work	Fun!	Work	Sleep

Figure 5.1. Weekly planner for a six-course schedule.

since your priorities lie in engineering study. Remember to schedule enough relaxation and leisure time. Your performance in engineering also will depend on your stress and anxiety levels, as well as how happy you are in your everyday life.

Scheduling your entire week might seem awkward and difficult at first, but persevere. You'll find it becomes easier the more you do it, and the benefits are enormous. Soon, your study time will become scheduled study time and reach *commitment status*. This is exactly what the very best engineering students do. They are quick to realize that study time is every bit as important as scheduled time, if not more important. Remember, study time is when you *process* the information *gathered* during scheduled time (e.g., in class, labs, etc.). Being an engineering student is very much like working for yourself: the benefits are directly proportional to the amount of effort expended in the appropriate directions.

Figure 5.2 contains a blank weekly planner similar to that shown in Figure 5.1. Why not take this opportunity to look at your weekly schedule. I'll bet there is significant room for improvement.

Make a List!

Making lists is a great way of ensuring that you accomplish what you need to accomplish. Whenever you have several things to do in any period of time, write them down in order of priority, and work your way through the list, *ticking off* each item as it is completed. For example, suppose you allocate two hours' exam review time to calculus. Your list for that period of time might look like this:

    i.   Go to Students Union and pick up sample exams.

    ii.   Look at textbook for explanation of Fundamental Theorem of Calculus.

    iii.   Review class notes on exam preparation.

    iv.   Work through first sample exam.

Figure 5.2. Weekly schedule.

	Monday	Tuesday	Wednesday	Thursday	Friday	Saturday	Sunday
0700							
0800							
0900							
1000							
1100							
1200							
1300							
1400							
1500							
1600							
1700							
1800							
1900							
2000							
2100							
2200							

Again, doing this beforehand means that the time allocated to study will be used entirely for study. Lists are easy to make and can be made anywhere, anytime. I make them in restaurants, on buses, in bed, or whenever anything occurs to me and anything resembling a piece of paper is at hand.

Save Yourself Time—It's Easy!

Almost everything you do as an undergraduate engineering student has been done before. The assignments, the lecture material, the experiments, and even the examinations have all been mastered and completed to perfection by other people ahead of you, either former engineering students or professors. This gives you an excellent opportunity to save time by *learning from what other people have done*. Consequently, recognize that years of experience and wisdom and the successes that result from it are contained in the very many resources available to you, for example, your textbooks, this book, your professors, your fellow students, professional engineers in industry, and many more. Modeling those who have achieved what you hope to achieve can save you time, pain, and effort. Consequently, whenever possible, instead of embarking on an analysis by trial and error, seek out information that will give you answers and the most effective way of doing things. It never ceases to amaze me how many students will try to solve technical engineering problems off the top of their heads, without consulting class notes, reading textbooks, or asking for help. This is wasteful and stressful. There is no need to reinvent the wheel, and no one expects you to! Just find out how it was done, and do the same!

5.2 PREPARING FOR AN ENGINEERING COURSE: MAKING SURE YOUR PREREQUISITE WORKS

Most engineering courses (for example, mathematics, physics, chemistry, engineering mechanics, etc.) are *cumulative*. This means that your performance in any particular engineering course depends heavily on certain skills developed in previous courses. For every engineering course, there is a specific set of *prerequisite skills* deemed to be essential by your instructor. These skills depend on the level and content of the course you are about to take, but they are usually few in number and almost always form only a small part of any formal prerequisite course. This makes them easy to review or practice. Consequently, to prepare effectively, it makes sense not to spend hours reviewing all previous material, but to identify and target those particular prerequisite skills. This forms the basis of an established procedure for effective preparation that doesn't take long (in fact, it takes only a few hours), but is known to significantly improve one's subsequent performance.

Step 1: Make Sure That You Know What Your Instructor Assumes You Know

The first step in preparing effectively for any engineering course is to talk to each professor (or senior students who have already been successful in the same courses) and identify the most essential functional prerequisite skills for each course. Ask each professor exactly what he or she *assumes* you know from the very beginning (i.e., ask for a description of what the professor regards as essential prerequisite skills for the course—in the first year, this is often different from what you have been told in high school), and then make it your objective to *know what the professor assumes you know*. This will ensure that both of you start off on a level playing field. For example, one course common to all first-year engineering students is introductory calculus. One of the first major topics in this course is concerned with the evaluation of the expression

$$f'(x) = \lim_{h \to 0} \frac{f(x+h) - f(x)}{h} \tag{5-1}$$

Here f is a given function of h is some constant. The *calculus component* of (5-1) consists of the limiting process and the application and interpretation of the quantity $f'(x)$, known as the first derivative of the function $f(x)$. The *precalculus component* (i.e., that which the instructor *assumes* you know) is everything to do with forming, simplifying, and manipulating the expression:

$$\frac{f(x+h) - f(x)}{h} \tag{5-2}$$

For example, let $f(x) = \sqrt{x+3}$. Then, before any calculus (i.e., the limiting process and interpretation of the function) can be applied, you must first know how to simplify and prepare expression (5-2) using precalculus techniques:

$$\begin{aligned} \frac{f(x+h) - f(x)}{h} &= \frac{\sqrt{x+h+3} - \sqrt{x+3}}{h} \\ &= \frac{1}{\sqrt{x+h+3} + \sqrt{x+3}} \end{aligned} \tag{5-3}$$

The expression is now in a form suitable for the application of the theory of limits—the calculus part. In arriving at expression (5-3), you have used prerequisite skills such as the theory of functions (forming $f(x + h)$ from the given $f(x)$), rationalizing the numerator of a rational expression (removing the square roots from the numerator of expression (5-2), necessary before you can apply limiting procedures), and algebraic simplification of the ensuing quantity, to arrive at the final form, expression (5-3). Your calculus instructor is concerned mainly with what happens after you get to expression (5-3), that is, evaluating the expression:

$$f'(x) = \lim_{h \to 0} \frac{1}{\sqrt{x+h+3} + \sqrt{x+3}} \tag{5-4}$$

The process by which you arrive at expression (5-4) (i.e., the *precalculus*) is *assumed* to be familiar and well known to you. There just isn't enough time to undertake any significant review in this area. Knowing precalculus *before* the course begins will certainly save you time, effort, and considerable stress, as well as allow you more time to concentrate on the important *new* material.

The ideas pertaining to the preceding mathematics example hold equally for *all* engineering courses, not just calculus. It is your responsibility to identify the prerequisite skills for all your courses.

Step 2: Once Those All-Important Prerequisite Skills Have Been Identified, Make Sure They Work

Once prerequisite skills are identified, it is important to make them *functional*. In other words, make sure that you have these skills at your fingertips and ready to go. Don't be influenced by how you scored in the prerequisite courses themselves. This can often lead to a false sense of security. Instead, make sure that you attain the required *fluency* in the necessary skills. Do this through practice. Go to an old textbook used in the pre-requisite course (if you don't have one, get one from the library or ask your professor to recommend a book) and perform a few sets of simple exercises in the relevant tech-niques. Do perhaps five problems for each technique or skill. This will have two major benefits:

  i.   You will achieve the required fluency.

  ii. You will *reawaken* your problem-solving skills. These exercises will be your warm-up exercises. It's simple, but effective. You won't believe how much eas-ier it will make the transition into your engineering course.

Why Do Highly Qualified Students Fail in First-Year Engineering?

Every year I address the entire class of (over 500) new engineering students and ask them to think about one of the introductory engineering courses they have just started. It may be beginning calculus, engineering mechanics, physics, chemistry, or something similar. I ask the students to raise their hands if they scored over 90 percent, 80 percent, etc., in the formal prerequisite course (usually a high school course). Most of the stu-dents indicate scores of above 70 percent, with a significant portion reporting scores of over 80 percent. Understandably, most of these students *assume* that they are indeed well prepared to move onto the next level (i.e., the present engineering course). I next ask the students to answer 10 multiple-choice questions in 20 minutes. These questions cover material from the corresponding prerequisite skills—the skills that I know have been assumed by the corresponding engineering professor. I have included the assess-ment test[1] I use when discussing beginning calculus in the following example.

ASSESSMENT TEST   The following test is designed to see if you are prepared for beginning calculus. Attempt all 10 questions. You should score somewhere in the range 8–10 (i.e., at least 80%) in approximately 20 minutes to consider yourself adequately prepared.

1. If $f(x) = \dfrac{1}{2}$ then $f(x + 2h) =$

   (a) $\dfrac{1}{2}(x + 2h)$ (b) $2h$ (c) $\dfrac{1}{2}(x + 2h)$ (d) $\dfrac{1}{2}$

2. The expression $f(x) = 2x^2 + x + 1, x \in R$ is
   (a) Always negative   (b) Always positive
   (c) Sometimes negative and sometimes positive   (d) Always zero

3. If $\dfrac{-3}{x+1} - \dfrac{2}{x-4} < 0$, then
   (a) $x \in (-1, 2)$ (b) $x \in (-1, 2) \cup (4, \infty)$ (c) $x \in (4, \infty)$
   (d) $x \in (-\infty, -1) \cup (2, 4)$

4. The expression $\cot x \sin^2 x \tan x \csc^2 x$ simplifes to
   (a) $\tan x$ (b) 1 (c) $\sin x$ (d) $\sec x \tan x$

5. Rationalizing the denominator in $\dfrac{1}{2 - \sqrt{3}}$ leads to
   (a) $\dfrac{1}{2 + \sqrt{3}}$ (b) $\sqrt{3} - 2$ (c) $2 + \sqrt{3}$ (d) $\dfrac{2 + \sqrt{3}}{4 - \sqrt{3}}$

[1] Reprinted with permission from *CALCULUS SOLUTIONS: How to Succeed in Calculus: From Essential Prerequi-sites to Practice Examinations;* Prentice-Hall Canada, ISBN 0-13-287475-X, 1997

6. $x^2 < 8x + 9$ is equivalent to

   (a) $-1 < x < 9$ (b) $x < -1$ or $x > 9$
   (c) $x < 1$ or $x > 8$ (d) $0 < x < 9$

7. If $x < 4$, $|x - 6| + |2x - 8| =$

   (a) $x + 2$ (b) $3x - 2$ (c) $4 - x$ (d) $14 - 3x$

8. The expression $4(2x+1)^{\frac{4}{5}}(2x-1)^{-\frac{1}{5}} - 8(2x-1)^{\frac{4}{5}}(2x+1)^{-\frac{1}{5}}$ can be factored to become

   (a) $-4(4x^2 - 1)^{-\frac{1}{5}}(3 - 2x)$ (b) $4(4x^2 - 1)^{-\frac{1}{5}}(3 - 2x)$

   (c) $8(2x - 1)^{-\frac{1}{5}}(2x+3)^{-\frac{4}{5}}$ (d) $8(4x^2 - 1)^{-\frac{1}{5}}$

9. The function $f(x) = \dfrac{1}{\sqrt{x^2 + x - 2}}$ has domain

   (a) $x \in (-\infty, -2) \cup (1, \infty)$ (b) $x \in (-\infty, -2] \cup [1, \infty)$ (c) $x \in (-2, 1)$
   (d) $x \in [-2, 1]$

10. If $f(x) = \dfrac{1}{x+1}$, the expression $\dfrac{f(x+h) - f(x)}{h}$, $h \neq 0$, is equal to

    (a) $\dfrac{1}{(x+1)(x+h+1)}$ (b) $-\dfrac{1}{(x+1)(x+h+1)}$

    (c) $\dfrac{-h}{(x+1)(x+h+1)}$ (d) $-\dfrac{1}{(x+1)^2}$

These questions deal with basic skills in factoring and simplification of algebraic expressions, solving inequalities, and reading and using trigonometric formulas, quadratic equations, functions, etc. After 20 minutes, I ask the students to pass their answer sheet to the person sitting in front of them, and we grade them together. In all the years that I have conducted this *experiment,* I have never had more than 1 percent of the class score a perfect 10. The majority score around 5 or 6 correct answers out of 10, with an alarmingly significant number scoring below 5. Many simply run out of time, while most have basically forgotten the relevant material. In either case, it is clear that the students' prerequisite skills are not as fluent and effective as they are required or assumed to be.

At this point, as you can imagine, I have the students' full attention. Perhaps most significant is the fact that many students are not even aware that they are ill prepared! Clearly, high averages in prerequisite (usually high school) courses often lull students into a false sense of security. In other cases, students have been absent from school for a significant period of time. Yet, because they have the formal *paper prerequisites,* they are admitted into engineering school, whereupon they do poorly because they lack prerequisite skills, or they have the skills, but they are nonfunctional. (Ask yourself this question: Would you fly in an aircraft piloted by a pilot who qualified for the position last year, but had had no air-time since then? More than likely, you wouldn't!)

Clearly, reacquainting yourself with prerequisite techniques is wasteful, is stressful, and takes time away from learning important new material. This is perhaps one of the main reasons that an excellent performance at one level (e.g., high school) does not necessarily translate into a similar performance at the subsequent level (e.g., engineering school) and that seemingly highly qualified students do not perform to expectations.

I always end my class by explaining what was just discussed to the new students: A paper prerequisite is worth only the value of the paper if it doesn't work. We discuss the two-step procedure previously mentioned, and I suggest sets of practice problems and examples (depending on the particular engineering course) to get the students' prerequisite skills back in working order.

Every year, many students return after their first year to let me know how this particular intervention turned things around for them. They also mention how, on applying these simple preparation techniques to all their engineering courses, from junior to senior level, they have seen significant improvement in their performance, each and every time. You should do the same, not just in your first year but in every year of your engineering education, particularly after the long summer between semesters.

5.3 EFFECTIVE NOTE TAKING

We have already discussed, in Chapter 3, the role of the lecture as a primary source of information in engineering school. In this section, we concentrate on how best to *record* that information so that you take as much as possible of the lecture away with you.

First of all, let's establish that there is absolutely no point in attending a lecture in engineering school, *unless you take notes.* Some students like to explain their inactivity during a lecture by saying that they learn better by listening rather than spending the whole hour or so writing things down. Here's what happens: In class, they probably understand most of what has been presented. One hour later, it's all been forgotten or replaced by the next lecture. Remember, lectures are your main source of information, information that is almost always used (at least) several weeks later. Your efficiency in a lecture is measured by how much relevant information you take away with you, not by how much information you understand at the time.

Effective note taking depends on several factors. Perhaps the most important of these is the ability to adapt to the many different ways in which information is presented in a lecture. Most professors have their own style of lecturing and communicating information. It is up to you to learn to adapt your note taking to suit the particular style of the professor in front of you.

In general, lecturers fall into three basic categories:

1. The lecturer that writes everything down. These lecturers are most common in technical courses, such as mathematics or science, where the *language of communication* is mainly mathematics and most examples, methods, and theories are presented with the intention that you copy them *word for word.* Your main responsibility in this type of lecture is to write down everything the professor writes (and the odd thing that he or she says, but doesn't write down). You are primarily gathering information; most of the thinking and understanding will come later, on your own time, when you review your notes.

2. The lecturer that writes down very little and instead prefers to *talk* his or her way through the lecture with the aid of, for example, an overhead and (many) different slides. This type of lecturer is more common in less technical courses (e.g., history, economics, sociology, english, etc.), where the information is more easily communicated using spoken English together with pictures, diagrams, and perhaps demonstrations. In these courses, it is usually very difficult to write down everything that the lecturer says: Most people cannot write as fast as the lecturer speaks. Consequently, it becomes necessary to learn how to isolate the most important information and record it most effectively for later review.

3. The lecturer that employs a combination of the previous two methods.

The following are suggestions that will help you improve your note-taking system. They are classified into the following three groups:

- *General remarks.* These suggestions are applicable to all three categories of lecturers.
- *Group 1.* These suggestions apply mainly to the first category of lecturers.
- *Group 2.* These suggestions apply mainly to the second category of lecturers.

Those of you with a Category 3 lecturer can combine the hints and suggestions given for Category 1 and 2 lecturers.

General Remarks

- Go to any lecture with the right attitude and objectives. The lecturer is there to help you—to give you the information you require to be successful.
- Don't talk while the lecturer is talking. Apart from the fact that it's rude, distracting (to the lecturer and to other students), and unfair to your fellow students, your conversation is likely being heard all over the room. Lecture rooms are specifically designed to carry sound.
- Take and keep notes in a loose-leaf binder. This has the following advantages:
  —It gives you the ability to insert and remove pages, in the appropriate places, as necessary. This is particularly important when the lecturer uses a combination of handouts and board or overhead materials, or when you wish to supplement or correct notes later upon reviewing them.
  —You don't need to carry around the whole semester's lecture notes with you all the time. Simply take the day's (new) lecture notes and insert them in the binder, at home, at the end of the day.
  —You can select the notes from any specific lecture(s) simply by removing them from the binder.
- Number, title, and date each page in the upper right-hand corner. This makes for easy reference and good organization.
- Leave a header at the top of each page, and use it to note any important information (e.g., midterm exam dates, assignments, etc.)
- Use standard paper and write on one side only! This will make it easier to supplement or edit notes later.
- Don't try to save paper. It may be good for the environment, but it will reduce the effectiveness of your note taking dramatically!
- Leave blanks for information you missed or for things that require further clarification. See the instructor after class or talk with fellow students, and then fill in the blanks.
- Organize your page appropriately: Separate topics, leave margins (for additional notes), and emphasize points using asterisks, uppercase letters, and underlining.
- Sit as close as possible to the front of the class, where there are fewer distractions and it's easier to see, hear, read the board, and identify important material.
- Focus: *Get into* the lecture. Tune into the lecturer while tuning everyone and everything else out. Leave your emotions outside the lecture room.
- Maintain eye contact with the lecturer whenever possible. Try not to doodle or play with your pen. Doing so will serve only to distract you.

- Get instructions on assignments and examinations *precisely*. Leave no room for doubt—ask if you're unsure. (Most assignments are usually given in lectures.)
- Never write your final notes in a lecture. As often as possible, use abbreviations and rough sketches and diagrams. This will minimize the effort you put into the actual writing, leaving you more time to concentrate on the information itself.
- *As soon as possible after the lecture,* review and complete your notes. Write them out as if you were teaching the subject to yourself. (Remember, this is exactly what you will be doing at exam time.) Leaving your lecture notes to one side until days or weeks after is just a waste of time and paper.
- If you are part of a study group, consider appointing a (rotating) note taker, who writes everything down while the remaining group members concentrate on the material being presented. After the lecture, get the group together and perfect a set of lecture notes for the group based on what three or more people (depending on the number in the group) have heard.
- When you miss a class, *always* get the notes from either a conscientious and successful student who shares your objectives in engineering study or, if possible, from the instructor. Use these notes as you would your own lecture notes (i.e., review, rewrite, supplement, and edit as necessary).
- If you want time to pass slowly, keep checking your watch. Otherwise don't check the time at all!

Category 1

- Try to prepare for the lecture by reading material assigned to the lecture *beforehand,* rereading the notes from the previous lecture, or *scanning* the material to be covered in the coming lecture from the textbook. You'll be amazed at how much more familiar lecture material appears if you've already thought about it or even just quickly read about it.
- Sit as close to the front of the class as possible, where you can easily see the board.
- Copy down *everything* on the board, regardless of whether or not it looks useful. Don't try to decide there and then what may not be relevant. Assume that if it's on the board, it's relevant. You will review your notes later anyway.
- Write down also anything the instructor might *say* (but not write down) by way of additional explanation.
- Note any references made to the textbook in the appropriate places, so you can look up an alternative explanation or more examples illustrating a challenging or difficult concept.
- After each lecture, if you haven't already, cross-reference your notes to the textbook. This will save you valuable time when you use your notes to complete an assignment or when you review for examinations.
- After each lecture, work through examples (presented in the lecture) *yourself,* and make a note of where (e.g., the textbook) other similar examples or the appropriate theory can be found.

Category 2

- Try to prepare for the lecture by reading material assigned to the lecture *beforehand,* rereading the notes from the previous lecture, or *scanning* the material to be covered in the coming lecture from the textbook or other sources of information. You'll be amazed at how much more familiar lecture material appears if you've already thought about it or even just quickly read about it.

- Do not try to take down everything the lecturer says. Most people just cannot write as fast as the lecturer speaks. Instead, spend more time listening and summarizing the main points in your own words: Not everything the lecturer says will be of equal importance.

- Listen *actively* by pausing to think before you write—but not for too long: You don't want to get too far behind.

- Write down the main ideas and any supporting details if possible.

- Don't let your own personal prejudices and biases affect your note taking. Be open minded and avoid arguing inside your head.

- Listen for any important clues as to what's important. For example, the professor might stress specific pieces of information, repeat certain points, change the tone of his or her voice, etc.

- Listen for words that signify an important idea or conclusion, such as "finally," "therefore," "consequently," "in conclusion," "to summarize," and so on.

- After the lecture, coordinate your reading and lecture notes, and expand on ideas and explanations as appropriate. Talk to the instructor and fellow students, if necessary, to reinforce ideas and understanding.

5.4 MAKING EFFECTIVE USE OF THE COURSE TEXTBOOK

Most engineering courses require that you buy a particular textbook. This textbook is chosen by your instructor to satisfy several requirements, including the following:

- To provide a backup source of course material. In this respect, the textbook will include the vast majority of the material covered in the corresponding engineering course.

- To provide a source of illustrative examples.

- To provide a good supply of practice problems.

It is important to understand that the textbook is intended to complement your engineering course and is in no way a substitute for lectures, labs, or seminars. To illustrate this point, consider the following situation. You are enrolled in *Engineering Mechanics I*. On the first day of classes, the professor gives you the following instructions: "There will be no lectures in this course. Everything you need to know is contained in the first five chapters, of the prescribed textbook. I suggest you buy the textbook, study the first five chapters and return to class on October 21 for the midterm examination and December 12 for the final examination." What kind of problems do you think you would encounter? Perhaps (at least) the following:

- The first five chapters of the textbook contain a wealth of information, but you can't hope to learn it all. What's important and what isn't? Which results can you omit without worrying about whether or not they will appear on the examinations? What are the main results of each section? Each chapter? There's just too much information.

- There are numerous practice exercises at the end of each section. You just do not have the time to work through every problem. Which of the problems are most important (e.g., for examination purposes)? Which of the problems are not relevant for this particular engineering mechanics course (i.e., different professors emphasize and include different topics)? Which of the problems are of examination quality? Which of the problems are similar to the type likely to be asked on a test?

Without the professor's guidance, it is extremely difficult to target relevant material from the textbook. One of the main reasons for attending lectures is that the professor has spent considerable time and effort targeting relevant material, examples, and practice problems. In lectures, the professor will tell you what is important and what isn't, what material may appear on an examination and what material is not likely to, which examples best illustrate the theory and which don't, and which practice problems will give you the most effective preparation for assignments, class tests, and examinations.

This is not to say that the textbook is not important. It is an integral part of any engineering course. However, there are ways to make sure that the textbook works *for you* in conjunction with the usual series of lectures, labs, and seminars.

- Use the lecture material as your *guide* to what's relevant. If you need a second opinion (or alternative explanation) or more detail on a particular result, consult the textbook. The alternative (or more detailed) explanation may be clearer to you.

- Again, with the class material as your guide, use the textbook as a source of more illustrative examples of the material covered in class. The more examples you see, the more likely it is that you will understand the basic concepts and that you will be able to apply the procedure to a new problem.

- Use the textbook as a supply of additional practice problems, but target problems *similar* to the ones solved in class or appearing on assignments or tests. This gives you smart practice, that is, relevant practice in the areas deemed to be important by your professor. In this respect, it is always a good idea to ask your professor to point out which problems in the textbook are most relevant.

- Many textbooks illustrate important theories and concepts using real-world examples and applications. These illustrations might help to make both theory and applications a little clearer to you. Hence, in addition to using the textbook as a source of problems and worked-out examples to help you solve homework problems and prepare for examinations, read through the accompanying text; it will enhance your understanding of the subject matter.

- Often, when a particular theory or concept is confusing, look ahead in the textbook to an application of the theory or concept in question. This might make things clearer for you.

5.5 HOW TO BE EFFECTIVE ON ASSIGNMENTS

Why are assignments (or homework problems) important? Certainly, one reason is that they can be worth anything from 5 percent to 20 percent of the final course grade. Another, perhaps more important, reason is illustrated in Figure 5.3.

The data in the figure represent average (final) examination scores for groups of students missing up to a maximum of 5 assignments in one of my engineering courses. It is interesting to note that the average score among those students submitting all 10 assignments was close to 70 percent, while that for those missing even 4 assignments was less than 50 percent.

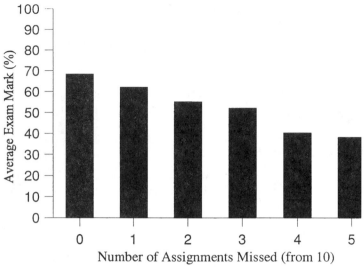

Figure 5.3. Missed assignments and their effect on final course grade.

It is clear that missing assignments also affect one's performance on an examination! To see why, we need only recall the basic principle that *we learn best by seeing examples and practicing for ourselves.* Assignments are vehicles for relevant and targeted practice. In other words:

- Assignments tell you *where* you should concentrate your practice. They identify relevant practice problems in relevant areas, as suggested by your professor!
- Assignments identify problems that best illustrate key concepts to enhance understanding.
- Assignments identify problems similar to those that may be asked on tests. (Your professor is giving you clues!)
- Assignments identify required standards and expectations. The problems chosen indicate the standard expected in a particular engineering course.
- In engineering courses, working through relevant practice problems allows you to see patterns and note repetitions in solution techniques, making subsequent problems easier to solve. Assignments provide an excellent source of such problems.

Now that we have established the importance of assignments, the next step is to examine how you can be most effective when *writing* assignments. As previously mentioned, the three main purposes of an assignment are:

- To get *practice* in relevant techniques.
- To get that 10 to 20 percent of the course grade allocated to assignments.
- To help you review for examinations.

All three of these objectives depend on how you *present* solutions to assignment problems—in other words, *what* you write and how it is *recorded.* To demonstrate, consider the following example. The subject matter is not important here, only the *procedure* and the *presentation.* That is, whatever is said with respect to this example is true for any problem in any engineering course. The example is as follows:

EXAMPLE 5.1.

A motorcycle can attain a maximum speed of 55m/s, accelerate at 5 m/s^2 and decelerate at 7 m/s^2. (s is seconds and m is meters). If the motorcycles starts from rest, find the minimum possible time it takes the motorcycle to stop exactly 800m further down the road.

SOLUTION

This problem is from an assignment dealing with *kinetics of a particle*. We begin by regarding the motorcycle as a particle.

- **We want:** the *time* it takes the motorcycle to travel 800m and stop, given the constraints in the question.
- **We know:** the motorcycle has a *constant* acceleration (5 m/s^2), a *constant* deceleration (7 m/s^2) and a maximum speed of 55m/s. Using the equations for position and velocity of a particle with constant acceleration

$$v = v_o + a_c t \tag{5-5}$$

$$d = d_o + v_o t + \frac{1}{2} a_c t^2 \tag{5-6}$$

where, here, v is velocity, t is time, d is distance, v_o is initial velocity, d_o is initial distance and a_c is the constant acceleration—everything is measured positive to the right starting from $d = d_o = 0$ and ending at $d = 800$.

- **Basic Strategy:** To travel 800m in the shortest time, the motorcycle should accelerate to its maximum speed and stay at that speed for the longest possible time i.e. making sure to leave enough distance to be able to decelerate to a dead stop by the time it has travelled 800m. Hence, the solution will consist of the sum of three times. The first time, t_1, will be the time it takes the motorcycle to accelerate from rest to its maximum speed of 55m/s. The second time, t_2, will be the time the motorcycle remains at that maximum speed. The third time, t_3, will be the time it takes the motorcycle to decelerate from its maximum speed to a stop at 800m.

1. *Time and Distance Required to Accelerate from Rest to Maximum Speed of 55m/s*

   From (5-5), since $v_o = 0$ (motorcyle starts from rest), $a_c = 5$ (constant acceleration), we have

$$v = v_o + a_c t$$
$$55 = 0 + 5t$$
$$t = t_1 = 11s \tag{5-7}$$

   Consequently, it takes 11s to reach the maximum speed. Using this value of $t = 11s$ in (5-6), since $d_o = 0$, the *distance* required to reach maximum speed is given by

$$d = d_o + v_o t + \frac{1}{2} a_c t^2$$

$$= 0 + (0)(11) + \frac{1}{2}(5)(11)^2$$

$$= 302.5m$$

2. *Time and Distance Required to Decelerate from Maximum Speed of 55m/s to a Dead Stop*

We repeat the process used in Step 1. previously except that now, $v = 0$ (our final speed is zero since we come to a complete stop), $v_0 = 55m/s$ (our 'initial' speed on deceleration is our maximum speed of 55 m/s) and $a_c = -7$ (our constant deceleration). Thus, (5-5) gives

$$v = v_o + a_c t$$

$$0 = 55 - 7t$$

$$t = \frac{55}{7}s$$

$$\simeq 7.86s$$

$$t_2 = 7.86s \tag{5-8}$$

Consequently, it takes $7.86s$ to decelerate to rest from the maximum speed. Using this value of $t = 7.86s$ in (5-6), setting $d_o = 0$, the *distance* required to decelerate to rest from maximum speed is given by

$$d = d_o + v_o t + \frac{1}{2}a_c t^2$$

$$= 0 + (55)(7.86) - \frac{1}{2}(7)(7.86)^2$$

$$= 216.07m$$

Thus, we have the following picture of the motorcycle's motion:

$\leftarrow$302.5m$\rightarrow$	$\leftarrow$281.43m$\rightarrow$	$\leftarrow$216.07m$\rightarrow$
$d = 0m$ ACCELERATION ZONE	REMAIN AT MAXIMUM SPEED	DECELERATION ZONE $d = 800$m

3. *Calculate the total time the motorcycle can travel at 55m/s.*

   From (5-6), using $d_o = a_c = 0$ (we're finding the time taken to travel the required 281.43m at a *constant speed of 55m/s*):

$$d = d_o + v_o t + \frac{1}{2}a_c t^2$$

$$281.43 = 0 + (55)t + \frac{1}{2}(0)t^2$$

$$t = \frac{281.43}{55}$$

$$\simeq 5.12s$$

$$t_3 = 5.12s \tag{5-9}$$

4. *Calculate the total time required*

   The total time taken to travel $800m$ in the shortest time is given by the sum of (5-7), (5-8) and (5-9):

$$t = t_1 + t_2 + t_3$$

$$= 11 + 7.86 + 5.12$$

$$= 23.98s$$

In this example, we have followed a clear, logical procedure in arriving at the correct answer. In addition, we have recorded all the steps in the calculation and the precise reasoning used to arrive at each step. Also, we have a record of exactly where each piece of subsidiary information can be found (in the course textbook or otherwise).

As far as the first purpose of an assignment is concerned (to get *practice* in relevant techniques), the procedure by which you arrive at the correct answer is just as important as the answer itself. It is exactly the development and repeated application of a set procedure that will allow you to begin to see patterns and repetitions in solutions, which is the main reason for practice in any engineering course. For example, if you were now asked to solve a related problem, you could mimic the solution to the preceding example. The reasoning is clear and easy to follow, because of the effort invested the first time around. In this respect, solutions to assignments should resemble *maps* or *recipes*, recording *where* and *how* to proceed when you want to repeat a similar calculation.

The following are additional reasons to record as much as possible of your solution procedure (bear in mind the other two purposes of an assignment: to get that 10 to 20 percent of the course grade allocated to assignments and to help you review for examinations):

- A good step-by-step solution procedure enhances understanding; it's like teaching or explaining the material to yourself.
- It may take some *hard thinking* to solve certain problems. Recording the steps will save time next time you encounter a similar exercise, leaving you more time to practice with new problems. (Recall the *map* concept, i.e., that maps basically allow us to draw from our own or someone else's experience of having already worked out how to get to a desired location.)
- The person grading the assignment will be impressed by your clear and logical thinking (indicating understanding) and will find it much easier to allocate partial credit when, for example, an arithmetic slip or a careless omission has led to an incorrect final answer.
- When it comes to tests (remember, assignment questions are excellent as practice test problems), a detailed approach to assignment solutions allows for *reviewing* as opposed to *relearning*. There's nothing worse than trying to remember how you did something and then having to spend the same time (again) *relearning* how to do it. Just keep a record, and then consult your *map* or *recipe*.
- Detailed solution procedures will help you organize your thoughts and develop effective problem-solving techniques. (See Chapter 7.)

There is no doubt that more detailed solutions will, at first, take more time. This extra time should be thought of as an investment in the future. Soon, the *step-by-step* approach will become automatic, and you will have developed an effective procedure for solving many different kinds of problems.

DOING ASSIGNMENTS YOURSELF VERSUS DOING THEM IN A GROUP

In Chapter 4, we discussed the benefits of collaborative learning. These benefits are particularly attractive when it comes to working on assignments in groups. We review some of them here for your convenience. Working on your assignment as part of a group will give you the following advantages:

- You will see and learn alternative problem-solving strategies.
- You will learn from your fellow students. This will add to and improve your overall learning experience.

- If you get stuck on a particular problem, the group tends to keep things going by *brainstorming* new and alternative suggestions instead of just giving up.
- The group atmosphere provides for active, cooperative learning: People in groups tend to verbalize their thinking, which adds to the problem-solving process.
- Working as part of a group will also lead to an improvement in self-esteem and a decrease in anxiety levels. (Talking about things relieves anxiety.)

However, when all is said and done (unless you are required to submit a group assignment), the final assignment you submit must reflect your *own* understanding—you will not have the benefit of a group in examinations. For this reason, it is perhaps best to adopt a middle-of-the-road approach. By all means, formulate and discuss ideas and opinions in collaboration with your colleagues. Think of this as part of the information-gathering process. Then use the information to put together your own assignment. Do not submit anything you do not understand, even if it is correct. First, make sure you understand how to solve all of the problems yourself, and then (with the three objectives of an assignment in mind) write and submit your own assignment.

5.6 USING POSTED SOLUTIONS TO ASSIGNMENTS

In many engineering courses, professors will provide detailed solutions to assignment problems. It is always a good idea to consult these solutions, irrespective of your grade on a particular assignment. There are two main reasons for doing so:

1. By consulting the professor's solutions, you see exactly what constitutes a proper, correct solution (procedure) in the eyes of the professor (usually the person who grades examinations). This way, you know exactly what's *expected* of you.

2. In some cases students achieve perfect scores on assignments, despite making errors and often not using the most effective solution procedures. In my experience, this is almost always attributed to the fact that teaching assistants grading assignments may not always examine every detail of each solution. Remember, an assignment is not just a way of accumulating points (recall the three purposes of assignments); it is also a way of developing efficient solution procedures and an important component of preparation for tests. For these reasons, whenever possible, you should always check your assignment against the professor's solutions, just to make sure that you are doing what you should be doing. When it comes to course examinations, it is almost always the professor who does the grading. I cannot tell you the number of times I have been approached by students (after I have graded their examinations) who ask me why points were deducted from a particular question when they "did it the same way in the assignment and got full credit!" Then I answer by reminding them why I post solutions to assignments.

5.7 USING TUTORS AND STUDY GUIDES

In Section 3.4, we discussed a variety of campus resources available to engineering students. In some universities, such resources include free tutoring services in, for example, writing skills, mathematics and learning strategies. You are strongly encouraged to make use of all free tutoring available to you, throughout your time in university, whether it is offered as part of your engineering program or as an additional campus resource.

In many cases, however, particularly when free tutoring is not available, engineering students choose to seek help by hiring a personal (or group) tutor or by purchasing self-study manuals or study guides. Whereas there are many advantages to using any of the latter, the biggest drawback, by far, is the additional cost. For this reason, *before* spending any additional money on help, it is always best to utilize, as much as possible, the sources of help available as part of your engineering program and those offered by campus organizations.

HIRING A TUTOR

The main reason for hiring a tutor should be to obtain *specific, targeted, intensive, one-on-one help,* at convenient, selected times during a course. The additional cost involved means that most people hire tutors only when it is absolutely necessary to do so, such as at critical times during the course, when an extra injection of intensive help is required (e.g., in preparing for examinations).

Choosing a tutor, however, can be fraught with difficulties. Let's begin by noting a few important facts concerning tutors.

A tutor is *not:*

- Someone who "sort of" remembers something about the material or was "once able" to do something similar.
- A student who has recently passed the same engineering course, even if he or she was at the top of the class! Passing a course (even with a perfect score) does not qualify someone to teach the material!

A tutor *is:*

- Someone with considerable expertise in the course in which help is required. A general rule of thumb is that the tutor should have completed at least two levels above that with which he or she is required to assist.
- Someone who can communicate ideas easily, effectively, and with a great deal of patience!
- Someone who can answer specific questions quickly, efficiently, and without any uncertainty.

The best way to find a tutor is through *word-of-mouth referral,* usually from your professor, a fellow student, or an informed advisor. Some institutions claim to have lists of *approved tutors.* Be careful here! Make sure you ascertain what qualifies a tutor for inclusion on the list. Sometimes it's nothing more than the fact that someone made a phone call to the person.

Once you have found a suitable tutor, it is necessary to take steps to ensure the most effective use of (*paid*) time spent in consultation. In this regard, the following suggestions may help you:

- *Prepare.* Before you meet with the tutor, make a list of all the questions you need to have answered. Keep the questions brief and specific. Make sure you are familiar with the material behind each question and the reason that you have asked a particular question. This will ensure that you make the time spent in consultation as effective as possible.
- *Control the session.* The tutor is there *for you.* Consequently, run the session according to *your* requirements. In this respect, make sure you tell the tutor

exactly what you want. Stick closely to your list of questions. Don't let the tutor *stray* from the topic, and don't waste time.

- *Stay focused.* Every minute of time spent with a tutor costs money. Hence, stay alert and focused on your requirements. Don't leave a particular point or question until you are entirely satisfied with the explanation.

- *Get targeted help.* Use the tutor as a *source of help,* not to replace your efforts on assignments and practice problems. The tutor should *add to,* not replace, the overall learning experience. The tutor cannot, and should not, do the work for you. Rather, he or she should *show you how* to do the work yourself.

- *Learn as part of a group if you can.* If you require less in the way of one-on-one attention, but would like to have access to a tutor for the purpose of discussing a certain number of specific points and questions (e.g., while reviewing or working through old examinations), consider getting together with your study group or a few classmates and hiring a tutor *for the group.* This will save you money and afford you the advantages of group discussion.

BUYING AND USING SELF-STUDY MANUALS

Self-study manuals are similar to tutors in that they can be used to provide *specific, targeted help* in relevant areas. Consequently, the choice of such a manual should be dictated by your specific requirements. For example, if you require a collection of worked-out past examination papers in introductory calculus, a book entitled *Understanding Mathematics* is probably less useful to you than one entitled *Calculus,* which itself is probably less useful to you than one entitled *Introductory Calculus,* which in turn is probably less useful to you than one entitled *Solved Practice Examinations in Introductory Calculus.* (See, for example, [4] and [5].) The key to making the most effective purchase lies in the *relevance* of the material to your requirements. Targeted information is quick, to the point, and usually less expensive than that contained in books addressing more general issues.

When buying a self-study manual or study aid, bear in mind the following:

- Choose a book that offers information closest to your requirements.
- Choose something that is relatively easy to read and explains things clearly.
- Your professor or someone else with experience in the area may be able to recommend a book that meets your specific requirements.
- Word-of-mouth referral is extremely effective. Ask what other people are using, have used, or would recommend.
- One of the biggest advantages of buying information in *book form* is that the information can be accessed almost anywhere, anytime, for example, in the bath, on the bus, in the library, in class, or at home.

Problems

1. Recall the last school day. Write down a schedule of how you spent your time on that day. Are there any ways in which you could have made better use of your time or increased your productivity? How? Rewrite the same schedule in a form that would have led to the best use of your time.

2. Using a form similar to that in Figure 5.2 and your answers to Problem 1, schedule your time for the next school day. Try to stick to this schedule, but feel free to make any minor adjustments as they arise. Explain any difficulties that arose. Did you find any advantages to scheduling your time?

3. Repeat Problem 2, but schedule your time for a whole week.

4. When you allocate study time, do you also write down where and what you will study? If not, how long does it take you to decide these matters? How much more study time could you have in a week if you decided *beforehand* where and what to study?

5. Take the assessment test in Section 5.2. What do your results tell you about your fluency in the basic prerequisite skills required for calculus?

6. Make up five-question tests (similar to that in the assessment test in Section 5.2) to test *prerequisite skills* required for each of your engineering courses. Use your own experience so far (which prerequisite skills did or do you think are most important for each of your engineering courses?) and any information offered by your professors or senior engineering students.

7. For each of your engineering courses, list five skills that are prerequisites for entering the course.

8. Write a one-page paper explaining the reasons one should attend lectures, and entitle it "Why I Go to Lectures."

9. List what you regard as the five main purposes of attending a lecture.

10. What do you think are the five main skills required to be effective in a lecture?

11. Take a look at the lecture notes from one of your most recent classes. Can you describe (in one or two sentences) the professor's objectives in that particular lecture? Show your answer to the professor who gave the lecture. Ask your professor if he or she agrees that these were indeed his or her objectives for that lecture. What does your professor's answer tell you about your effectiveness in recording information from lectures?

12. List five things that you can do to improve your effectiveness in a lecture. Implement these suggestions in the next lecture you attend, and describe how they affected your performance.

13. What is your main source of information in each of your engineering courses. List four other sources of information. Rank your list from 1 (most valuable source) to 5 (least valuable source).

14. Explain the importance of taking notes in a one-page paper entitled "Why Bother Taking Notes in a Lecture?"

15. Why do you think engineering professors prescribe or recommend textbooks? List five reasons.

16. List five ways in which you use your textbook. Rank these from 1 (most frequent use) to 5 (least frequent use).

17. Suppose you were given a choice of using lectures or a textbook? List five advantages associated with not having to attend lectures and having to rely on the textbook instead. List five disadvantages associated with not having to attend lectures and having to rely on the textbook instead.

18. Write a one-page paper on the importance of assignments and homework. List the reasons that professors assign homework and what you think homework adds to your overall performance in engineering courses.

19. Look over a problem from one of your previous assignments, preferably from two or more weeks ago. Pick one of the more difficult questions. Read over your solution to this problem. Does the solution still explain how to solve the problem today? If not, improve the solution so that in two weeks time you can read it through and understand immediately how to solve the problem.

20. What part do assignment questions play in preparing for examinations? List five advantages of using assignment questions to prepare for course examinations.

21. Does reading through the solution to a problem provide the same level of understanding as solving the problem yourself? Explain. List five differences between the two procedures.

22. Suppose you get stumped on a particularly difficult question about impulse and momentum from your engineering physics course. One of your study buddies has solved the problem and gives you a copy of his solution. What do you do?

23. Of the study strategies presented in this chapter, list those in which you actively engage. List those that you have never tried before. List those that you think would help your effectiveness as an engineering student. Resolve to implement as many of these strategies as possible throughout your time in engineering school.

24. So far in this book we have discussed the following study skills and strategies:

  - Interacting with, and making effective use of, your professors
  - Collaborative learning and group study
  - Time management
  - Making the most of class time: lectures, laboratories, and seminars
  - Effective note taking
  - Goal setting
  - Getting yourself motivated
  - Making use of campus and off-campus resources
  - Preparing for your engineering courses: making your prerequisites work
  - Making effective use of the course textbook
  - How to be effective on assignments

Which of these strategies do you think are most important for success in engineering study? Rank your choices from 1 (most important) to 11 (least important). Which skills or strategies do you need to improve in? Devise a plan for improving these skills.

6

How to Be Successful on Examinations

Examinations invariably make up the single largest contribution to your final grade in any engineering course.

This simple fact explains why many engineering students are focused on examinations (quizzes, midterms, and finals), rather than on an appreciation of the course material. Like it or not, your performance on examinations will more or less determine how well you do in your engineering courses. For this reason, it is essential to understand *why* the very best engineering students are so successful in examinations and to learn how to use this information to your advantage.

So why do certain students perform so much better than others on examinations? Some students put it down to a simple matter of intelligence: "Oh, that girl is really smart. Her father is a physics teacher and her mother has a Ph.D. No wonder she scores over 90 percent on all her tests." Others put it down to the lack of a social life: "That guy never goes out. He does nothing but study. No wonder he performs so well on tests!" I suggest that neither is entirely correct and that the truth lies somewhere in between. At senior high school and junior college levels, *intelligence* alone is no longer sufficient to place someone in the top 5

OBJECTIVES

In this chapter you will learn how to:

- Prepare for tests.
- Be successful on tests.
- Write answers to test questions and obtain maximum points.
- Overcome exam anxiety.
- Recover if you fail a test.

percent of the class. There is far too much material to absorb and not enough time in which to absorb it—even if you devote all your time to studying. In fact, it has been my experience that the very best students have an extremely active social life. Indeed their level of success often increases with their level of activity. Achieving success has more to do with *how* you prepare for an examination and *what* you do to prepare.

I learned early on in my academic career that

> knowing the course material ≠ success in course examinations.

I recall my midterm examination from my first engineering mechanics course. I had worked consistently throughout the year, understanding the course material, doing every assignment, and working through extra practice problems (in much the same way as I did when I was in high school). I understood the main ideas and concepts, and I was able to apply them in different situations. So I was quite confident that I would do well on the examination. Imagine my surprise when I discovered that I had scored only 58 percent! Worse than that, many of those students scoring above me had performed poorly throughout the course having missed assignments and often asking *me* for help. I couldn't understand why this happened. I had worked hard and I knew the material, so why wasn't I performing to the best of my ability? I began to discover the answer to my question when I asked one of my classmates (who had the highest score on the test) how she had prepared for the examination. It became clear to me that there were some *missing ingredients* in my test preparation routine. Basically, it came down to two things:

- Smart practice
- Examination technique

My friend and I had both prepared well during the course. What made the difference in our midterm scores was what we did in preparing for the test itself. She had obviously regarded the test as a separate entity, targeting and tailoring all her efforts not solely toward reviewing the course material (as I did), but toward doing well *on the test itself* (*smart practice*). She had obtained many former and practice midterm examinations and rehearsed her performance, so that she had a much better idea of what was expected and how to demonstrate the required knowledge under a time constraint (*examination technique*). She was entirely focused on doing well on the examination. I, on the other hand, was focused on the course material, believing that to be sufficient to perform to the best of my ability on the midterm.

To understand why my friend's strategy was so much more effective than mine, let's return to the car-driving analogy. None of us believe that we can pass a standard driving test simply by driving the way we do in everyday life. We recognize that a driving test requires us to demonstrate a distinct collection of maneuvers and exercises, based on basic driving skills, under examination conditions. Conversely, no one continues to drive the way they did during their driving test. The latter is a rehearsed performance, requiring specific targeted practice based on a knowledge of exact requirements (*smart practice*) and a focused effort to perform well under specific test conditions (*examination technique*). Consequently, in preparing for a driving test, we find out as much as we can about *what* is required and *target* our preparation (as effectively as possible) toward those particular goals. Exactly the same principles apply to preparing for any test, academic or otherwise.

In fact, since that first mechanics midterm, an acknowledgment of these basic principles has allowed me to perform to the best of my abilities on all subsequent examinations, such as academic examinations (in many different disciplines), driving tests, athletics competitions, or whatever requires me to demonstrate performance under a given set of constraints.

In this chapter, we discuss, in detail, the many different aspects of maximizing performance on examinations, including, in particular, the two main ingredients: *smart practice* and *examination technique.*

6.1 PREPARING FOR EXAMINATIONS: SMART PRACTICE AND EXAMINATION TECHNIQUE

To prepare effectively for course examinations, it is important to think about *both* long- and short-term preparation. Each is an essential component of an effective overall exam-preparation strategy. Long-term preparation builds a solid base or foundation in the course material, allowing time for learning, understanding, and asking questions. The material is slowly digested and assimilated into long-term memory, where it can be recalled relatively simply (with, for example, a few practice problems) and as required.

Short-term preparation is more concerned with the details of the examination itself and involves fine-tuning, targeted (or smart) practice and practice in ones examination technique.

There are many examples in real life where this type of long- and short-term strategy is employed. For example, when training for competition, athletes build a solid base throughout the year using weight training and a combination of speed and endurance exercises (long-term preparation). Only as competition time approaches, do they tend to devote most of their training to *specific* events (short-term preparation).

LONG-TERM PREPARATION

Clearly, what you do during the semester will affect your performance on course examinations. As previously noted, long-term preparation is not, in itself, sufficient to guarantee maximum performance on examinations, but it is certainly necessary. Those who omit the long-term component of exam preparation engage in what is more commonly known as *cramming.*

Cramming describes an effort to fit a significant amount of work (for example, a month's or an entire semester's worth of work) into a single night or weekend. *Cramming never works in engineering,* basically because most disciplines contributing to the subject are not fact based. Rather, they are method based, cumulative (the understanding of one part depends heavily on the understanding of previous parts), and example driven. There is just no way to read, understand, and practice every technique in a such a short period of time. It simply doesn't work. Engineering students who try cramming for examinations *never* perform to the best of their abilities.

Fortunately, long-term preparation takes place automatically for those who work consistently and conscientiously throughout the semester. The main components are summarized as follows for your convenience:

- *Lectures and classroom time*
  —Collect information in a clear, concise, and organized fashion. This will make the information easier to understand when it comes to reviewing for examinations. (See Chapter 5.)
  —Pay particular attention to specific topics and sections of the course material and examples emphasized in class. They often unexpectedly show up on examinations.

—Make notes of any hints or extra information that a professor might give during the course.

—Use class and lecture materials to identify relevant sets of practice problems from the textbook or any other source. These will be useful when it comes to practicing, or *fine-tuning*, for examinations

- *Assignments and homework*

  —Write clear, concise, and logical solutions. This will help tremendously when it comes to reviewing for examinations, since it is always easier to *review* than to *relearn*. (See Chapter 5.)

  —If you answered any assignment problems incorrectly during the semester, make sure you understand your mistakes and make a note of the correct solution(s). It is always best to do this as soon after realizing the error as possible, when the topic is current and fresh in your mind. When you return to these solutions (for example, in reviewing assignments), not only do you have the correct solution but you also have a note of what *not to do*—that is, your original error.

  —Use assigned problems as an indication of what material the professor deems most important in the course and of the standard expected by the professor. I often tell my students to think of assignments as a collection of *study guides* for each particular topic covered in the course.

- *Developing efficient problem-solving techniques.* Using clear and logical procedures to solve a wide range of problems over the entire semester cannot help but develop efficient problem-solving techniques. These techniques improve only with experience and practice, neither of which cannot be achieved in the short term. By examination time, many of these techniques will become *automatic,* requiring only fine-tuning when you review for examinations.

- *Asking for help.* Fix problems as they arise (not all at once near exam time, which leads to panic and *exam anxiety*). Use all of the resources at your disposal, including your instructor. (See Chapter 3.) Make a note of the help you receive, because you may need it later if you encounter similar difficulties upon reviewing course material.

The most successful students continue to demonstrate the importance of consistent long-term effort as a necessary foundation on which effective exam-preparation strategies are based. There is no doubt that this has always been, and will continue to be, an extremely significant factor in distinguishing optimum from mediocre performance on course examinations.

SHORT-TERM PREPARATION

Short-term preparation is mostly concerned with preparing for the examination itself and involves fine-tuning, *targeted* (smart) practice, and practice in examination technique. The following are the main components of an effective short-term strategy:

- *Find out what will be covered on the examination.* It is always extremely useful to know which material is likely to be on an examination and which isn't. This allows you to target your efforts toward the relevant areas. There are basically two ways to get such information:

  a. *From (recent) past examinations in the same course.* In many introductory engineering courses, examinations tend to cover the same standard material, year after year. Use this information to your advantage. Note any patterns, relative emphasis of one topic as opposed to another, and, perhaps most important, which topics appear most often. (Shortly, we shall discuss how and where to obtain old or practice examinations.)

b. *From your professor.* Ask your professor what material will be covered on the examination. This is a perfectly valid question, and you have nothing to lose by asking it. (But remember to be polite, courteous, and professional.) In most cases, the professor makes up the examinations, so there is no better person to ask. Similarly, when you peruse old examinations, show them to your professor. Ask whether the old practice examination is a good example of what will be on the [professor's] exam. Ask the professor if he or she recommends that you work through any *specific* past examinations or set of practice problems.

- *Apply smart practice.* Once you have a good idea of the topics most likely to appear on the examination, concentrate and target your efforts toward these particular areas by using the following steps:

1. Review the relevant theory from course notes and the course textbook.

2. Review relevant worked-out examples from the course notes and the course textbook.

3. Review your solved assignments.

4. Work through relevant problems:

   i. *Practice problems.* Work out sets of relevant practice problems from the textbook or any other source (preferably suggested by your professor). Choose these practice problems carefully: The main purpose here is to develop fluency and a working knowledge of selected techniques. Consequently, the practice problems should be mainly repetitive. Once you feel comfortable with any particular method, try a different style of problem using the same technique, but perhaps in a different setting. (These types of problem usually appear towards the end of a particular problem set in the textbook.) If you are unsure whether a particular set of problems is relevant, *ask* your professor. A good rule of thumb is to perform at least five (standard) practice problems per technique (to develop fluency), followed by one or two *different* or unusual problems in the same technique (for fine-tuning).

   ii. *Assigned problems.* Redoing assigned problems is excellent practice, since you have the correct solution in front of you against which to check your work. Also, assigned problems are good indicators of the required standard and are chosen specifically to reinforce relevant lecture material that is likely to appear on examination.

   iii. *Suggested problems.* Your professor may have worked through or suggested specific problems in class while a particular topic was under discussion. Do these problems (again if necessary), for they have been chosen specifically to reinforce or demonstrate a particular method or technique. Finally, *ask* your instructor to recommend some practice problems. You can be sure that this information will be relevant.

   iv. *Old or past examinations.* This is discussed in detail next.

- *Work through old or practice examinations.* Without a doubt, this is the most crucial stage of *smart practice.* An old or practice examination affords you the opportunity to actually rehearse the event. The questions are as relevant as they can be (they are actual examination questions for the very same engineering course), and the examination conditions are probably identical, or at least extremely similar, to those that you will encounter. There are two equally important components of an old or practice examination package:

a. *The examinations themselves.* These can be obtained from many different sources:

    i. Your professor

    ii. Departmental offices

    iii. The campus bookstore

    iv. Exam registries in the Student's Union.

    v. Students who have taken the course previously.

    It is always a good idea to ask your professor which old or practice examinations are most relevant and which he or she would recommend.

b. *Detailed solutions to the examinations.* These are usually extremely difficult to obtain, but get them when you can. They are extremely valuable. Not only do they allow you to check your own solutions, but they let you see what is expected of you. Sometimes, professors will make solutions available, and sometimes the solutions are sold in packages in departmental offices or bookstores. It may take a little effort to find them, but it is always worthwhile: They make a particular examination much more effective as a tool for practice or rehearsal. Be careful, however, to use the solutions properly. Don't read them as a substitute for working through the problems. (*Reading the solution $\neq$ doing the problem yourself.*) The solutions will always look easier than expected, and there is no substitute for performing the procedure yourself. You should pretend that you don't have the solutions, struggle with the problems as necessary (this is where the majority of the learning takes place), and consult the solutions only after you feel you have finished. Remember, there will be no solutions available when you take the examination. If you cannot get solutions to a particular practice examination, try the examination yourself, and then ask a member of the teaching staff to help you with any difficulties or to check your solution technique for obvious mistakes. Remember, the correct answer is only part of the solution; the technique by which you arrive at the answer is just as important. (See Section 6.3.)

There are two ways to use old or practice examinations:

1. *As an excellent source of relevant practice problems.* Ignore the examination conditions, and just do the problems. This will develop the required fluency and fine-tune your skills.

2. *As a way to develop your examination technique.* When you take into account the actual examination conditions, practice examinations allow you to rehearse, while actual old examinations are equivalent to a dress rehearsal; you can actually simulate the examination itself. Simulating the examination will allow you to develop the skills required to perform under the constraints (time, stress, or otherwise) of an actual examination.

    I always recommend to my students that they should work through at least three practice examinations before writing their particular examination.

At least one of the three examinations should be an actual past examination, and they should treat it as a *dress rehearsal.*

- *Examination technique.* Examination technique is concerned with two things: *acknowledging* that an examination requires you to perform under certain constraints and *practicing,* as much as possible, to overcome any difficulties associated with those constraints (for example, exam anxiety and time management). The following are the constraining elements inherent in most examinations:

  —Examinations always incorporate an unknown element: You are never 100% sure of what you may be asked.

  —Examinations require that you perform under a time constraint.

  —Examinations require that you demonstrate your knowledge precisely and logically. This means that, to be most effective, you must present your solution in a manner compatible with that expected by the person grading the examination. We shall discuss the different aspects of writing an examination, including the most efficient way to present solutions and what the person grading the examination looks for, in Section 6.3.

  Taken together, these constraints are largely responsible for the two most common complaints associated with writing examinations: *exam anxiety* and *insufficient time.*

  —*Exam anxiety.* It has been my experience that the most successful engineering students overcome exam anxiety by making the examination an *anticlimax.* By the time they get to the examination, they have worked through so many practice problems and practice examinations, that they are on "autopilot." They know what to expect, many of their reactions during the examination are instinctive, and they are focused on their particular goal. There will always be the usual adrenaline rush associated with writing an examination, but when it comes to overcoming exam anxiety, *preparation* is the key—particularly short-term preparation, in the form of *rehearsal* and *dress rehearsal.* I have found that the process of working through practice examinations is one of the most significant components of overcoming exam anxiety in engineering.

  —*Insufficient time.* This is again symptomatic of inadequate practice and rehearsal with actual examinations. Working through an adequate number of practice problems allows you to develop fluency in the necessary skills, making your approach and problem-solving procedure almost automatic. This means that, in the examination, you solve problems quickly and effectively. Practice or old examinations, on the other hand, tell you what to expect in the allotted time. Working through a sufficient number of practice examinations cannot help but inform you of how much material you are likely to encounter in the examination itself. You will know (approximately) what to expect, and you can practice performing the required number of problems in the allotted time. We will return to this particular topic later, in Section 6.3, when we discuss writing the examination.

It is clear that old or practice examinations play an important role in developing one's *examination technique.* Other important factors that determine your effectiveness when you take an examination also will be discussed in Section 6.3.

6.2 PREPARING FOR EXAMINATIONS: GETTING ORGANIZED

As an engineering professor, I have always believed in going from the specific to the general. My approach to this particular chapter is no different. In Section 6.1, we discussed the specifics of how to study for a test: *smart practice* and *examination technique.* In this section, we take one step back and examine the more general issues related to exam preparation.

How you prepare for an examination depends very much on the particular type of examination you are required to take. For this reason, it is important to obtain as much general information as possible about the examination itself, as soon as it becomes available. The following are things to find out about the examination:

- *Details of the examination*
    —Is it a quiz, midterm, or a final examination?

    —What are the place, time, and duration of the examination?

    —Is it an open-book or a closed book examination?

    —Is it a multiple choice, a written examination, or both?

    —Are *"cheat sheets"* (formula sheets) allowed?

    —Are calculators allowed?

    —How much is this examination worth as a percentage of the final grade?

    —What happens if you miss the examination for any reason?

- *What might appear on the examination?* Once you have discovered the type of test facing you, you should then enter your *short-term* preparation routine, as previously discussed. The first step in this routine is to find out what is likely to be asked on the test and what isn't. This will begin the process of *smart practice,* as discussed before.

We have already considered, in detail (Chapters 3 to 5), some of the things you should do as part of an effective long-term preparation strategy. The following are additional suggestions that may help you during your short-term preparation:

- *Prepare a review schedule.* Prepare a structured review schedule. This schedule will vary in length and depth, depending on the volume of material you have chosen to study. This, in turn, will depend on the type of test you are required to write. For a final examination, prepare your schedule at least two weeks in advance. For a midterm examination, prepare one to two weeks in advance. For a quiz, preparing a few days in advance often will suffice. When preparing your schedule:
    —Scan the relevant course material, and divide it into separate sections, usually based on different theories, techniques, or applications.

    —Identify the material that you think is likely to appear on the examination.

    —Allocate study time to each section according to the volume of material, its relevance to the examination, and its importance (i.e., the likelihood that it will appear on the examination).

    —For each section, identify a set of relevant practice problems from the textbook, course notes, past assignments, or otherwise. Make note of these.

—Decide which practice or old examinations you will work through and when you will do it. (Remember to keep one such examination for the final stages of your review as a *dress rehearsal.*)

—Remember to allocate more time for particularly difficult concepts or examples that require more thought such as word problems. (See Chapter 7.)

• *Review section by section.*

—Begin with an overview of the technique illustrated in a particular section.

—Read through worked-out examples which use that technique (e.g., from course notes or your textbook) until you feel confident enough to be able to apply the technique for yourself (i.e., until you understand the main ideas behind the method.

—Begin practicing the technique for yourself using the sets of (targeted) practice exercises identified earlier. Be sure to write clear, logical, and *methodical* solutions to the problems, as if you were teaching the material to yourself. (See Section 5.5.) This will help you pick up extra points when you actually take the examination. (See Section 6.3.)

—Once you feel comfortable with the technique in a particular section, move on to the corresponding practice problems from course notes and old assignments. You should have access to the solutions to these problems, so make sure you check your *method* as well as your answer. (The former is more important; see Section 6.3.)

—By this stage, you should have a good working knowledge of the section of material you are reviewing. To *fine-tune* your skills, pick a problem or two (dealing with this material) from some recent or practice examinations, and see if you can confidently and competently write a full solution to each problem. Do not skip any steps. Get into the habit now of writing full comprehensive solutions. Remember, when you take the examination, you must demonstrate your knowledge. Don't assume that any particular step is trivial; it may not be to the person grading the examination. (See Section 6.3 for more on this.)

—Finally, to complete the review of a section, write a summary of the section as follows:

List the important concepts, techniques, and formulas in the section.

Link each concept, technique, and formula with practice problems, assignment problems, examples in course notes or the textbook, and practice examination problems that you have worked through as part of your review and have found to be particularly good for understanding and developing fluency in the technique. You may want to return to these problems and examples for a quick review of the section as the examination draws closer.

Make any notes that you think may help you when you return to this material later.

—Follow the foregoing procedure for each section of examinable material.

• *Study groups: discussion.* Group review is particularly effective for the following reasons:

Group review represents

—active, cooperative learning. It is always a good idea to *talk* about the material under review.

—You see and learn problem-solving strategies.

—You learn from other students.

—When you get stuck, the group will tend to keep the momentum going.

—Group review improves self-esteem and decreases anxiety levels.

Consequently, *in addition to* (not instead of) following these review procedures, make an effort to discuss the material, examples, and problems with other people. Be careful to use this activity properly (as a *supplement* to your review). I have witnessed many situations in which students working in groups believed that they were reviewing effectively, but were, in fact, merely taking notes from the efforts of the well-organized students who always led the discussions. (See the discussion on group work in Chapter 4.) When all is said and done, you will face the examination alone. It pays to keep this at the back of your mind at all times!

- *Final review and fine-tuning.* Following the suggestions discussed above will allow you to learn the material, develop fluency and method in the particular techniques, and commit all of this to a part of your brain from which it can be easily retrieved. Then, one or two days before the examination you can review the collection of summaries you made when you reviewed each individual section. Build confidence and reinforce what you already know by selecting random practice problems from your collection and solving them *"blindly"* (i.e., without any supplementary information, such as that obtained from notes or a textbook, as you will do in the examination). Finally, fine-tune your skills by *rehearsing* with one or two complete examinations. After this, you will be well prepared for anything!

In closing this section, we make a few common sense suggestions relating to the logistics of studying.

- *Where to study.* To study effectively, it is necessary to free yourself of distractions and competing associations. I have found over and over again that the thinking part of the brain really *warms up* only after a period of deep thought or effort, usually after struggling or trying very hard to solve a particular problem. For this reason, you should never mix business with pleasure when it comes to studying for examinations. Instead, you should try to make your study time as efficient and as effective as possible. In my experience, one hour of concentrated, focused study is worth three hours of watered-down, distracted study any day. Try to keep the following simple rules in mind:

  —*Pick a quiet room that is free of distractions.* For example, the bedroom or library, and *not* your bed, the living room of your home, the kitchen table, a local fast-food restaurant, or a cafeteria or coffee shop.

  —*Get comfortable.* Make sure that when you are studying, you wear comfortable clothes and use a comfortable chair. Any discomfort will distract you from your main purpose.

  —*Take frequent breaks.* Be sure to get out of your study environment for a break whenever you feel the need, perhaps every hour or so. This will keep you sharp and maximize your effectiveness. Beware, however: Breaks shouldn't occur too frequently. It usually takes at least 30 minutes of effort to warm up your thinking. You shouldn't interrupt your concentration just when you get into things. Breaking too frequently will mean that you are constantly warming up and never working at the most effective level. Keep the breaks short (e.g., five minutes) and simple. Get up for a stretch, a snack, or something else, but be careful to minimize distractions during your break. Keep things rolling over in your mind, and don't get into some deep (unrelated) conversation with a friend!

—*Eat well and get plenty of rest.* To perform well mentally, you need to stay healthy. If you organize your study time effectively, you will have sufficient time to eat well and get lots of sleep.

—*Engage in physical activity.* For some reason, after spending a significant period of time thinking about a problem, things often come to me in the most unusual places or when I'm doing something completely unrelated, such as jogging or working out at the local gym. I'm not sure why, but the complete change in activity (from studying) seems to make things clearer. Many of my friends and colleagues appear to have had similar experiences. Have you ever had that experience? Even if you haven't, some good physical activity does tend to refresh and reenergize our minds and our bodies. And it doesn't have to be anything sophisticated: Even taking a brisk walk seems to have the desired effect.

6.3 TAKING THE EXAMINATION

The day has arrived! You're well prepared, confident, and ready to go. Nevertheless, to maximize your performance during the examination, there are certain essential components of taking an exam of which you must be aware.

- *Eat something.* Make sure you have a good meal on the day of the examination. You will expend lots of energy when you take the examination, so make sure you fuel your body sufficiently.
- *Dress comfortably.* Wear comfortable clothes. Remember, you may be sitting in the same position for up to three hours.
- *Do you have everything?* Before you leave for the examination room, go through a checklist of all the things you will need when you take the examination. For example, you may need any or all of the following things:
  —Writing instruments
    Pens
    Pencils
    Rulers
    Erasers
  —A calculator. If you do need one remember to check the batteries. Also bring spares if possible.
  —A textbook or other supplementary materials allowed by the examiner (if you will be taking an open-book examination)
  —A watch

  Bring as many replacements of these items as you think you will need.

GET THERE EARLY Arrive at the examination site at least 15 minutes before the examination begins. This will give you time to compose yourself, note the seating plan, and make yourself aware of any new instructions.

GET A GOOD SEAT When you enter the examination room, make sure you choose a good seat, one that is relatively free of distraction. In some cases, large rooms are used for a variety of different examinations, of varying styles and duration. If your examination is two hours long, and the row next to yours is being used for a one-hour examination, you will be distracted by students packing up midway through *your* examination. In such cases, it always pays to take a few minutes before the examination to study the seating plan. Similarly, try not to sit near students with heavy colds, they tend to sniff and cough a lot. Also students (even friends) with a different examination philosophy (i.e., those that tend not to take examinations seriously) should be avoided when it comes to seating. You have invested too much time and energy to risk being distracted by someone who is not focused. Find a nice quiet area not too far from the people proctoring the examination, because you may need to ask questions.

The large majority of examinations in engineering are of the written type, multiple-choice or an element of each. We begin with a discussion of the written examination and return to the subject of multiple-choice tests later in the chapter.

MAXIMIZING PERFORMANCE IN WRITTEN EXAMINATIONS

The First Few Minutes   When the examination begins, spend the first few minutes scanning the questions (including the distribution of points). When you do, note (in writing near, each question) which technique you will use and how much time you think you will need to answer each question. These little notes help you allocate your time effectively and act like doors, opening compartments to the (now) more-than-familiar corresponding review sections. Scanning the entire test at the start also gives you an opportunity to make sure that your examination paper is complete. Imagine finding out with five minutes to go, that you are missing a question worth 20 percent of the grade! Once you have looked at all the questions, rank them according to level of difficulty. The easier problems usually require you to demonstrate methods and set procedures. (See Chapter 7, Type A problems.) The more difficult problems require more thought and less routine application of course material. (See Chapter 7, Type B problems.)

At this point, you can proceed in one of two ways, depending on your particular preference:

1. Start with the easier problems and work towards the more difficult problems. There are three main advantages to this strategy:
    i. In solving the easier problems, you slowly *warm up* your thinking in preparation for the more difficult ones.
    ii. Solving a series of problems successfully means that you gain immediate confidence that you can tackle the more challenging problems.
    iii. Getting the easier problems out of the way first will maximize the amount of time remaining to consider the more challenging problems.
2. Start with the most difficult problem and work towards the easiest problem. The main advantage in using this approach is that you can tackle the questions requiring the most thought at the beginning of the examination, when you are less tired and more alert.

In my own engineering courses, the majority of the top students have consistently favored the first approach, which is also the method I used as a student, but it depends on your particular preference. Once things are underway, bear in mind the following points, which will add to your overall effectiveness when you take the examination:

- *Note the point distribution.* Use the point distribution to allocate time for each question. Clearly, a question worth five points should not require as much time as one worth 20. Use the point distribution also as an indication of what is expected: The 20-point question requires 20 points worth of effort, and so on.

- *Show details in open-book examinations.* Open-book examinations rely less on memory and more on method and technique. Consequently, you are expected to demonstrate more detail in open-book examinations than in closed-book examinations. For example, in a closed-book examination, you may get a point or two for writing down a correct formula. In an open-book examination, by contrast, the formula is considered as being *supplied* (in the text), so it carries no weight. Instead, open-book examinations emphasize more method and problem-solving techniques.

- *Use formula sheets.* If a formula sheet is supplied with your examination, use it as a guide to the techniques that are to be employed on the examination. For example, if a complicated formula does not appear in the formula sheet, it is unlikely that you will need to use that formula on the examination.

- *Attempt every question.* Don't be afraid to try to answer every question, even if you're not sure how to proceed. Write down your thoughts, and try to develop a solution using logical steps. Partial credit may be awarded for some of the things you write down.

- *Watch your time.* Pace yourself, and try to stick as closely as possible to the time you have allocated to each question (on your initial appraisal of the examination). If you get stuck and can't seem to make any progress on a particular problem within the allocated time, leave the problem, and return to it at the end of the examination if there is time. It is better to lose points on one problem and gain points on the remaining problems than to sit for the rest of the examination wasting valuable time. Remember, even if you have been engaged in active thought in answering a particular question, if there is nothing on paper, the instructor will assume that you have done nothing in that time. There is no way for the instructor to believe otherwise.

- *Ask for clarification.* If you are unsure about anything to do with how the examination is written, including the wording of a particular problem or the way it is stated, *ask* about the problem. There is nothing to lose, and you gain the added advantage that a verbal clarification might jog your memory.

- *Write with the purpose of getting the maximum number of points.* This is perhaps the most important aspect of writing examinations. An examination is just that—an *examination* or *investigation* of your performance, in a particular subject, on a particular day. Accordingly, you are required to *perform,* that is, to demonstrate your knowledge of the subject. When you bear in mind that in a written examination the only way to demonstrate ability or knowledge is by writing it down, your answers you will understand that the person grading the examination will use the written response as the *sole* criterion for judging your ability to answer a particular question. This is the crucial consideration that must be taken into account when writing solutions to a problem or an exam. Solutions must be written with the person grading the examination in mind.

After I give an examination, some students always return to discuss their performance. Some wish to see their paper in an effort to pick up more points. What follows are some of the explanations I have heard whenever students have tried to explain blank or partial solutions to examination problems:

- "I knew what I was doing; I just didn't write it down!"
- "You [meaning me, the professor] know that I know how to do this stuff! I've done it many times in the assignments, and you gave me 100% each time. Surely you didn't expect me to produce all the detail. There is a time limit, you know!"
- "I worked out the problem on a piece of scrap paper. I didn't write the details down because I didn't think they were important. I did get the correct answer, however; look!" (The student then points to the one equation $x = 3$ on the otherwise blank page.) "Don't I deserve full credit for this problem?"
- "I didn't have time to write the complete solution, so I did the calculation quickly in my head and obtained the correct answer. Why did you give me only 1 point out of 10?"

I answer these comments by informing the students (yet again; I do this in class at least three or four times *before* the examination) of exactly what I (and instructors, in general) look for when grading examinations:

- A set procedure or method illustrating clear, logical thinking and understanding, leading to the correct answer
- An ability to use the most appropriate technique in the most efficient way possible
- An ability to communicate ideas effectively
- An ability to develop a scientific argument, stage-by-stage, step-by-step, and with any necessary mathematics, leading to the desired result
- An ability to use problem-solving strategies and explain the significance of any results obtained

Clearly, the final answer (e.g., $x = 3$) is only one part of the *solution;* accordingly, it carries only a proportion of the points allocated to a *complete* solution.

In Section 5.5, we discussed the importance of writing clear, logical, step-by-step solutions to assignment problems. Exactly the same is true of writing solutions to examination problems, except that in this case it is even more crucial. Your professor will equate what you have on paper to what you know about solving a particular problem. There is nothing else to look at.

Below are three different solutions to the same problem taken from a final examination in Engineering Mechanics: Statics. The first two solutions were submitted by Students #1 and #2 as part of the written examination for this course. Both solutions have been graded (using the solution key provided in Solution #3—the professor's solution) and show the number of points awarded for each correct step in the solution.

PROBLEM   A bullet-proof vest consists of a series of identical thin coated metal sheets fixed together. A bullet, moving at 900ft/s, is directed in a straight line towards the vest. The bullet loses 20% of its speed as it passes through the first sheet in the vest. How many such sheets should the vest contain to ensure that the bullet stops before passing completely through the vest? (Neglect air resistance and assume that the bullet's mass is conserved).

SOLUTION
(Student #1)

$$T_1 + work = T_2$$
$$p\ sheets$$

$$\frac{1}{2}\frac{W}{g}(900)^2$$

$$\overset{?}{\bullet}\ \frac{1}{2}\frac{W}{g}(720)^2 \quad \text{①}$$

$$\int F\,da$$

$$\text{⑤/⑩} \qquad p = \frac{145800\,\dfrac{W}{g}}{\dfrac{1}{2}\dfrac{W}{g}(900)^2} = 2.78 \checkmark$$

$$\sim\ \underline{3\ \text{sheets required}} \checkmark \quad \text{②}$$

PROBLEM | A bullet-proof vest consists of a series of identical thin coated metal sheets fixed together. A bullet, moving at 900ft/s, is directed in a straight line towards the vest. The bullet loses 20% of its speed as it passes through the first sheet in the vest. How many such sheets should the vest contain to ensure that the bullet stops before passing completely through the vest? (Neglect air resistance and assume that the bullet's mass is conserved).

SOLUTION #2
(Student # 2)

$$\frac{1}{2}mv_1^2 + W.D. = \frac{1}{2}mv_2^2$$

Particle dynamics

Work + E   ②

$$\frac{1}{2}m(900)^2 \qquad\qquad \frac{1}{2}m\left(900 - \frac{900}{5}\right)^2$$

$$\frac{1}{2}m(900)^2 + N\left[\frac{1}{2}m(900)^2 - \frac{1}{2}m(720)^2\right]$$

⑤/⑩

$$= 0 \qquad ②$$

$$\underline{N = 3}\checkmark \quad ①$$

PROBLEM | A bullet-proof vest consists of a series of identical thin coated metal sheets fixed together. A bullet, moving at 900ft/s, is directed in a straight line towards the vest. The bullet loses 20% of its speed as it passes through the first sheet in the vest. How many such sheets should the vest contain to ensure that the bullet stops before passing completely through the vest? (Neglect air resistance and assume that the bullet's mass is conserved).

SOLUTION #3
(Professor)

- **Strategy:** Regard the bullet as a particle. The bullet is moving with an initial speed which changes after it comes into contact with the first metal sheet. In other words, the bullet's kinetic energy is altered as a result of the sheet. Try principle of work and energy.

- **Principle of work and energy.** Consider the situation arising when the bullet passes through the first sheet.

$$T_1 + Work(1 \to 2) = T_2 \qquad (1)$$

Here, T_1, T_2 represent the kinetic energies of the bullet just before and just after it passes through the sheet, respectively.   ②

- **Calculate kinetic energies and work**

$$T_1 = \frac{1}{2}mv_1^2 = \frac{1}{2}m(900)^2$$

$$T_2 = \frac{1}{2}mv_2^2 = \frac{1}{2}m(720)^2 \qquad (2)$$

where m is the mass of the bullet and v_1, $v_2 (= \frac{4}{5}v_1)$ are the speeds of the bullet before and after impact, respectively. Before impact, the sheet does no work on the bullet. As the bullet passes through the sheet, however, the sheet does work equal to $-I$ on the bullet (since the force exerted by the sheet on the bullet is in the direction opposite to the bullet's motion). Hence the total work done by the sheet on the bullet from time $1 \to$ time 2 is

$$0 - I = -I \qquad ② \qquad (3)$$

- **Apply principle of work and energy for a single sheet.** From (1) - (3),

$$\frac{1}{2}m(900)^2 + (0 - I) = \frac{1}{2}m(720)^2$$

Solve for I:

$$I = 145800m \qquad ② \qquad (4)$$

- **Apply principle of work and energy for the required number of sheets.** Suppose we require n sheets to stop the bullet. Noting that, in this case, since the bullet is stopped in the final state,

$$T_2 = 0$$

From (1), we obtain

$$\frac{1}{2}m(900)^2 + (0 - nI) = 0$$

From (4), we have:

$$\frac{1}{2}m(900)^2 + [0 - n(145800m)] = 0 \qquad ②$$

- **Solve for *n* to obtain**

$$n = \frac{25}{9}$$
$$\simeq 2.78$$

②

In other words, at least three sheets are required to stop the bullet.

Notice that, although both Student #1 and Student #2 arrive at the correct answers, neither demonstrates the (complete) correct logical procedures required to arrive at these answers (illustrated in the professor's solution.) Consequently, both score only 5 points out of a possible 10, even though they each obtained the correct answers. They may have *known* the correct procedures, and they may even have *thought* through the problem using these procedures, but the point is that they did not *write* down the details. The grader cannot be expected to read between the lines or work out what you are trying to say. Only what you write down will be used towards assessing your grade. Solutions are almost always graded on a partial-credit basis, meaning that points are allocated to different parts of the procedure, of which the correct answer is but one part. To get a perfect score, you must demonstrate *all* of the components of the procedure. This grading methodology also works *for* you, when, for example, you demonstrate the correct procedure, but arrive at the wrong answer through an incorrectly performed calculation or some other misstep. In this case, you will receive all the points except those allocated to the correct answer itself. Try to apply the following guidelines when you write solutions in examinations:

- Recognize what the instructor will look for when grading your solution.
- Develop a clear, logical procedure leading to the correct answer.
- Don't be messy! The person grading your exam will not take the time to decipher what you are trying to say.
- Don't assume that the person grading the exam knows what you are talking about. Tell the grader exactly what you want to say.
- Label diagrams and place them where they belong, near the part of the solution to which they are most relevant.

Finally,

- *Don't Leave Early*
  Devote the entire examination period to maximizing your performance. If you finish the examination early, use the extra time to check your solutions. Add details, tidy up explanations, or think about alternative strategies for dealing with problems you couldn't solve. Even if you think you have aced the test, stay and check anyway.

- *Ignore Everyone Else*
  Sometimes, when fellow students leave the examination early, there is a tendency to think that the examination should be easy and that you are missing something. This is nonsense. Take all the time you need. For all you know, those people leaving early may have failed. In fact, in my experience, examinations submitted ahead of time almost never account for the top scores in the class. Quite the contrary, the very best students use every available minute to their advantage.

MAXIMIZING PERFORMANCE IN MULTIPLE-CHOICE TESTS

Most of what has just been said of written tests is true also of multiple-choice tests. After all, both types of test have the same objective: to test your knowledge of the course material. From your point of view, the main difference and the biggest drawback to a multiple-choice test is that, since each question requires only a simple answer, for example, A, B, C, or D, there is little opportunity to *demonstrate* understanding of the course material through the presentation of well-written, clear, and logical solutions. Consequently, unlike written tests, in which method and procedure contribute significantly to the number of points awarded for each solution, multiple-choice tests exclude the opportunity to accumulate points from a detailed and methodical solution.

Nevertheless, in engineering, the approach to solving problems on a multiple-choice test should, essentially, be the same as that used in solving problems on a written test. The main difference is that, at the end of the procedure, only the final answer is returned, not the solution (the *method* leading to the final answer). In other words, in a multiple-choice test, you will be answering the same types of questions you would on a written test, only shorter ones (a consequence of asking many equally weighted questions in a relatively short period of time). Hence, you should be prepared to work through each problem *anyway,* on a separate sheet of paper, and *arrive* at the correct answer through the same logical reasoning used to solve problems on a written test. Accordingly, make sure you bring a supply of *scrap paper*!

Of course, in a multiple-choice test, since the correct answer appears alongside each question (although hidden among decoys), there is always the opportunity to *guess.* This method, however, should be used only as a last resort. By far the most efficient and effective way of tackling a multiple-choice test in an engineering subject is to treat it, essentially, as a written test and return only the final answers (at the end of your solution) on the answer sheet.

The following procedure may help you in answering multiple-choice questions:

Step 1 Identify any key words to help get you through the *padding* (extraneous information) and to the real meaning of the question. Identify the appropriate technique or theory to be used in the solution.

Step 2 Ignore the (given) answers and solve the problem for yourself on a separate sheet of paper. Do so clearly and methodically, as if you were supplying the solution to the same problem on a written test. If necessary, draw a diagram, it may help you to see things more clearly.

Step 3 Compare your answers with the given alternatives. If your answer is among them, enter the appropriate choice on the answer sheet. If your answer is not among the alternatives but *close* to one, go back and check how you worked out your solution. (This will be easy to do if you have solved the problem methodically.) Locate and correct any errors you find, and repeat each step until you believed you have identified the correct alternative. If, despite your best efforts, you cannot arrive at any of the given alternatives, choose the one that is *closest* to your answer.

Step 4 Move on to the next question.

The following tips may prove helpful in answering multiple-choice questions:

- If you are completely *stumped* and unable even to begin to solve a particular problem, eliminate any alternatives that you *know* are incorrect, and make an *informed guess* from the remaining alternatives.

- The solution to a multiple-choice question is usually short and should take no more than 5 to 10 minutes (a consequence of asking many equally weighted questions in a relatively short period of time). If you find yourself spending more than this amount of time on a particular question, you are probably using the wrong method, so either guess or move on to the next question. Remember, each problem is worth the same number of points.

- If you complete the test with time remaining, return to those questions that you answered with any degree of uncertainty. Reexamine the *logical alternatives* (i.e., exclude the alternatives previously identified as incorrect), and check your answers.

- Often, *working backwards* is a good way to check your answer. Take your answer and make sure it fits the problem.

6.4 AFTER THE EXAMINATION

After the examination, you should take any opportunity to view your graded paper. For class quizzes and midterms, this is usually done in class, on the day the professor returns the grades. For final examinations, it may be necessary to make a separate appointment with the professor. No matter how well you did, reviewing your examination is a learning experience and a vitally important exercise in preparing you for the next examination, in the same course or otherwise. The following are some of the more important advantages associated with reviewing your examination:

- *Identification of errors in grading.* The professor may have made an arithmetic error in adding the different points awarded to each problem. Check the addition yourself. Similarly, the professor may have missed a question or forgotten to grade part of a problem. Point this out if he or she did. Ask the professor to outline his or her grading scheme. This way, you can ascertain the various points allocated to each solution and ensure that your paper was graded accordingly. Remember to be professional and courteous at all times.

- *Learning from your mistakes.* Use what you have learned from reviewing your examination to improve your performance in any future examinations. While the material is fresh in your mind, check any posted solutions, try to understand any errors in your solutions, and rework problems as necessary. This is particularly important in reviewing quizzes and midterms' where you have the opportunity to correct any errors in method, understanding, and presentation (writing) *before* the all-important final examination. Consult with your professor as necessary.

- *Self-critique.* An examination affords you the opportunity to *try out* your studying, preparation, and examination-writing strategies. Use the review to analyze, improve, develop, and fine-tune your approach. Look for weaknesses, and decide how you will not make the same mistakes again.

- *Learning from the examination.* In the case of a quiz or midterm, you now have an actual examination made up by the very professor that will be responsible (entirely or in part) for your final examination. Note the style, the emphasis, and, above all, what the professor requires in a perfect solution. Be sure to use this information to your advantage in the final examination. Quizzes and midterms also afford you the opportunity to have your instructor critique and comment on all aspects of your performance, *before* the final examination. What better way to improve your studying, preparation, and examination-writing strategies than to have the *examiner* comment on the results of your efforts!

WHAT TO DO IF YOU FAILED MISERABLY

If you really made a mess of the examination and you're not sure why, apart from doing the things previously mentioned, you should make an appointment with your professor and try to explain or discuss what went wrong. The professor may see errors in your studying, preparation, or examination-writing strategies that may be easy to correct. I make a point of meeting with students who find themselves in this position, and I encounter the same recurring reasons that they have performed so badly on examinations:

- Most admit to never having worked through a past or practice examination *by themselves.*
- Most have never engaged in smart practice. (See Section 6.1.)
- Most have never thought about *rehearsing* and practicing their examination technique.

In most cases, it is the *short-term preparation* that is inadequate or, in some cases, missing entirely. This is easily corrected using the procedures outlined in Section 6.1.

It is important that you maintain adequate communication with your professor in this and in all course-related matters. At the end of the course, the professor is charged with the responsibility of assessing your performance using the usual sources of information (assignments, examination scores, etc.) and any other information that you care to make available. Thus, you can make a good impression on your professor by frequently asking sensible questions in a courteous and professional manner. In this way, you appear conscientious and committed. Consulting with your professor after an examination adds to that impression and works to your advantage. For example, in some cases, professors will discard a disastrous midterm grade and move the appropriate weight to the final examination. This usually happens only when the professor is convinced that you have the potential to demonstrate the required knowledge by the end of the course, but, for example, you have "gotten off to a bad start." Such a decision is almost always *ad hoc,* based on the professor's *impression* of your individual *demonstrated* abilities, a significant part of which depends on good communication between you and your professor.

Problems

1. Write a one-page paper entitled, "Why Examinations Are an Essential Part of Undergraduate Engineering," and discuss reasons this is so.
2. Write a 500-word essay entitled, "What I Do to Prepare for Examinations" in which you discuss your examination preparation procedures.
3. Think about your examination preparation routine. Write down its main components, in chronological order. What are its main strengths and weaknesses? Suggest ways to improve the weaker aspects.
4. Using the information in this chapter, devise a new examination preparation strategy. Implement this strategy when you prepare for your next test.
5. Write down five advantages of using past or practice examinations as part of your exam preparation.
6. List three places on campus where you can obtain past or practice examinations for each of your engineering courses. Find out if solutions are available.
7. How would you find out which topics are likely to appear on an upcoming test?
8. Produce a detailed *study schedule* for any one of your engineering courses. You should allocate sufficient time for studying and demonstrate how and when you will cover the necessary material, as well as perform the necessary practice examples.

9. Which of the 3 solutions presented on pp. 99–102 most resembles your solution to an examination problem? How could you improve on the way in which you present solutions in examinations?

10. Ask each of your professors how they go about grading examinations. Use this information to help you maximize your effectiveness when you take an examination.

11. Have you ever had a really bad examination? Explain why it was so bad and what you did or will do to ensure that you don't make the same mistakes again.

12. Make up an examination question from any one of your engineering courses. Try to be as original as possible. Produce the detailed solution and allocate partial credit accordingly. Make sure you test all the necessary skills the examinee should possess.

13. Make up a one-hour examination for any one of your engineering courses. You should produce a full solution key and grading scheme. You should also ensure that the best students are adequately challenged and that the time allotted is neither too little nor too great.

14. From your experiences answering Problems 12 and 13, write a one-page paper on "What a Professor Looks for When Grading Examinations."

15. Having read the material in this chapter, describe how your approach to preparing for and taking examinations will change. If you feel that there is no need for you to change anything, say why.

16. Describe a (nonacademic) real-life example in which you have prepared, taken, and successfully passed a test of some kind. Identify the most important factors leading to your success. Draw parallels with engineering examinations.

17. Write a one-page paper detailing what you would do if you discovered that you had failed an examination that you expected to *ace*. Include a discussion relating to how you would use the experience to improve your subsequent performance on tests.

18. Have you ever failed an examination? What went wrong? What did you change about your examination preparation and writing routines to prevent the same thing from happening again?

7

Procedures for Effective Problem Solving

We have already noted that examples and practice problems are essential components of maximizing one's performance in any engineering course. Consequently, a significant amount of course time (lectures, tutorials, labs, and assignments) is devoted to worked-out examples and relevant practice problems. In Chapters 5 and 6, we emphasized the importance of presentation in developing effective problem-solving techniques—in other words, how an effective solution requires that you demonstrate a clear, logical, and organized procedure. In this chapter, we examine actual problem-solving strategies. We illustrate our ideas with worked-out examples from different engineering courses.

Basically, two types of problems are encountered in engineering courses:

Type A   *Those that require mainly the application of known techniques and minimal thinking.* In other words, these problems require you to repeat from memory, know the meanings of certain key concepts, and apply established course material to new situations. Such problems are also known as plug-and-chug problems, solved by applying a formula or set procedure. This type of problem is common in introductory mathematics courses, such as beginning calculus, where you are often asked to use set procedures to, for example, differentiate . . . , integrate . . . , solve . . . , and so on.

SECTIONS

- 7.1 Solving Problems That Require Mainly Application: Type A
- 7.2 Solving Problems of Type B: Word Problems

OBJECTIVES

In this chapter you will learn:

- How to translate engineering problems into mathematics.
- How to solve word problems.
- How to write clear and logical solutions to engineering problems.
- The difference between *reviewing* and *relearning* engineering material.
- How to evaluate answers to engineering problems.

Type B *Those that require mainly thinking and minimal application of established techniques.* These problems exercise the higher level thinking skills and, as such, are often more difficult than Type A problems. They usually involve some *mathematical modeling,* followed by the *evaluation* and *application* of selected mathematical techniques (usually from the course material) and, finally, the *interpretation* of results in the context of the physical problem. Type B problems include those commonly referred to as *word problems* and are usually found in courses such as engineering mechanics and physics, where real physical situations must first be *translated* into mathematical language before known procedures can be applied.

In the sections that follow, we examine procedures used to solve each type of problem.

7.1 SOLVING PROBLEMS THAT REQUIRE MAINLY APPLICATION: TYPE A

Problems of Type A serve mainly to reinforce lecture material. They are chosen to encourage *practice* in the relevant techniques, to *actually apply the techniques.* In engineering, understanding the material is only one part of the learning process. The other (more important) part is concerned with actually applying the material by yourself in practice problems. This part is the one that contributes most to the learning process.

BLOOM'S TAXONOMY OF EDUCATIONAL
OBJECTIVES (COGNITIVE DOMAIN)

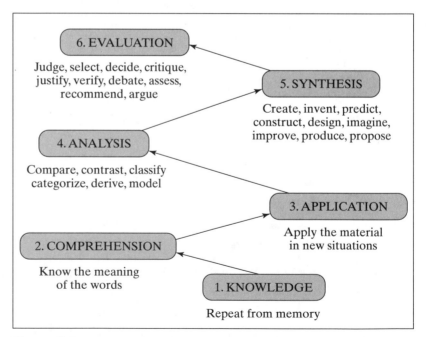

Figure 7.1. Levels 4–6 are the *higher level thinking skills.* Usually, undergraduate education deals almost exclusively with Levels 1–3. Excerpted with permission from R. M. Felder and J. E. Stice, *National Effective Teaching Institute Manual,* Anaheim, 1995

Reading through and understanding a well-written solution to a problem (from, for example, the textbook, class notes, or someone else's assignment) is not the same as actually solving the problem by yourself!

For example, how many of us learn to swim (or drive) simply by watching someone else and *understanding* what they are doing? That's the easy part. The tough part is getting into the water (or behind the wheel) and performing the movements for ourselves. Try the following exercise:

1. Find a (difficult) worked-out example, preferably a *word problem*.
2. Read through the solution until you understand exactly what's going on.
3. Cover up the solution (leave the problem), and place a blank sheet of paper in front of you.
4. Now solve the problem for yourself.

You'll quickly realize that solving a problem starting from a blank sheet of paper (which is exactly what you need to learn to do) is completely different from reading through the solution to the same problem.

Something else to remember about problems of *Type A* is that, since they are chosen primarily to practice a particular technique, by their very nature they belong to distinct classes (each class associated with the technique being practiced), each of which identifies a particular solution procedure. Hence, the first step in an effective solution procedure for problems of *Type A* involves identifying the class to which a particular problem belongs. For example, all problems involving finding the solutions of a quadratic equation follow a set procedure that makes use of the well-known formula for solving quadratic equations. Similarly, problems involving relative motion follow a set procedure that makes use of the equations for relative motion, and so on.

The following is a systematic problem-solving procedure that is extremely effective for solving *Type A* problems. To begin, imagine that you have an assignment or practice problem in front of you. Then:

1. Read through the problem carefully, and classify the problem by identifying the (general) corresponding area discussed in your class notes or textbook. For example, does the question deal with solving algebraic equations or differentiating functions? Or perhaps it deals with the kinematics of a particle or the conservation of energy. Whichever it may be, there will be a known framework or set procedure for solving the problem. Your job is to *find* and *mimic* that procedure. In classifying a problem that deals with more than one area, classify it according to the *primary* area. For example, a problem dealing with the equation of a tangent line to a curve, represented by a given function, may be chosen primarily for practice in the techniques of differentiation, even though some analytic geometry is used in arriving at the solution.

2. Once the general area has been identified, the classification has to be more specific. Read through the problem again, and decide what (exactly) you are required to find. For example, if you are dealing with a problem in matrices (in a course in linear algebra), are you finding the inverse or applying matrix multiplication procedures? If the problem deals with the kinematics of a particle (in a physics course), are you being asked to find the velocity or acceleration of the particle or the path the particle follows over a certain time? Get to the *heart of the problem*, and identify the specific area, technique, formula, or rule that will form the *basis* of the solution.

3. Once you know what you are looking for, go to your major source of information. This may be your class notes, textbook, or whatever is being used as your main source of material. Look for the corresponding area and, more specifically, that dealing with the main formula, rule, or technique that must be applied. Next, find a worked-out example similar to the one you are trying to solve, using that same formula, rule, or technique. This is an extremely important part of the problem-solving process. There is no need for you to *"reinvent the wheel";* all you need to do is adapt a successful solution procedure to your particular example. That's the nice thing about Type A problems: Most have already been solved, and those that haven't are similar to those that have. (Remember the main reason for solving Type A problems).

4. Apply the known procedure to your specific example, following each step carefully. Remember to be clear and logical and to explain (to yourself) at each step exactly what you have done. (See Section 5.5 to remind yourself of the reasons why this is important.)

Now you have the frame and most of the body of the solution, and your attention is focused on a specific area. If the problem is straightforward, you can proceed to finish the solution. Occasionally, you will encounter difficulties (and get *stuck*), perhaps because the problem you are solving has a peculiar characteristic that makes it slightly different from the example you are following. This is okay; in fact, it's an important part of the learning process. If you find yourself in such a situation, *don't give up!* Try to get around the problem by examining in a bit more detail the theory surrounding the example from the class notes or textbook. You will find that the harder you try, the deeper your understanding of the problem becomes. If, after spending some time thinking about things, you still cannot make any progress, get some help (from your study group, graduate teaching assistants, or your professor). Often, all it takes is a little nudge in the right direction. In the large majority of cases in which I help students with problem solving, I usually need spend no more than a few minutes with each student. The difficulty is almost always overcome on the first examination of the student's (partial) solution. In fact, most of the time, the student, still actively thinking about the problem, will come up with the answer during the discussion. That's why it's important to use help as a part of the problem-solving process and not as the problem-solving process itself. Seeking help prematurely is extremely counterproductive.

In the following examples, we illustrate the procedure just given using problems chosen from two typical first-year engineering courses, introductory calculus and engineering physics:

EXAMPLE 7.1. Differentiate the following function of x.

$$f(x) = \frac{\sin x \cos x}{x + 1}$$

SOLUTION
Preliminary thoughts:

1. General area: Differentiation of functions
2. More specifically: A quotient of functions but numerator involves a product of functions. Expect to use both quotient and product rules for differentiation and possibly trig. identities and factoring techniques.

3. Find a similar example in textbook or class notes—one that demonstrates the quotient rule.

4. Mimic that example to produce the correct solution as follows.

$$f(x) = \frac{\sin x \tan x}{x + 1}$$

First use quotient rule

$$f'(x) = \frac{(x + 1)\dfrac{d}{dx}(\sin x \tan x) - \sin x\left(\tan x \dfrac{d}{dx}(x + 1)\right)}{(x + 1)^2}$$

$$= \frac{(x + 1)\dfrac{d}{dx}(\sin x \tan x) - (\sin x \tan x)(1)}{(x + 1)^2}$$

To obtain $\dfrac{d}{dx}$ (sin x tan x), use the product rule for differentiation.

$$\frac{d}{dx}(\sin x \tan x) = \sin x \frac{d}{dx}(\tan x) + \tan x \frac{d}{dx}(\sin x)$$

$$= (\sin x)(\sec^2 x) + \tan x \cos x$$

$$= (\sin x)(\sec^2 x) + \sin x$$

$$= \sin x(\sec^2 x + 1)$$

Finally,

$$f'(x) = \frac{(x + 1)\sin x(\sec^2 x + 1) - (\sin x \tan x)(1)}{(x + 1)^2}$$

$$= \frac{(x + 1)\sin x(\sec^2 x + 1) - \sin x \tan x}{(x + 1)^2}$$

$$= \frac{\sin x[(x + 1)(\sec^2 x + 1) - \tan x]}{(x + 1^2)}$$

Excerpted with permission from "How to Study Mathematics," by P. Schiavone, Prentice-Hall Canada, 1998.

EXAMPLE 7.2.   A particle travels counterclockwise along a circular path of radius r = 50 meters with a constant speed of v = 15 meters/sec. Find the particle's angular velocity and the magnitude of its acceleration.

SOLUTION
Preliminary thoughts:

1. General area: Velocity and acceleration of a particle.
2. More specifically: Circular motion. Expect to use specific formulas for velocity and acceleration of a particle in circular motion—using polar coordinates.
3. Find a similar example in textbook or class notes—one that demonstrates the use of the same formulas.

4.   Mimic that example to produce the correct solution as follows.

Let e_θ and e_r be the basis vectors for polar coordinates. For a particle in circular motion, the velocity v is given by

$$v = r\dot{\theta}\,e_\theta \tag{1}$$

where $\dot{\theta}$ is the magnitude of the particle's angular velocity. We are given that $v = |v| = 15$ and that $r = 50$. Hence, from (1),

$$15 = 50(\dot{\theta})$$

$$\dot{\theta} = \frac{3}{10}\text{ rad/sec} \tag{2}$$

The particle's angular velocity is therefore $\frac{3}{10}$ rad/sec, counterclockwise. Similarly, For a particle in circular motion, the acceleration a is given by

$$a = -r\dot{\theta}^2 e_r + r\ddot{\theta}\,e_\theta \tag{3}$$

Since the particle's speed is constant for all time, $\ddot{\theta} = 0$. Hence, from (2) and (3),

$$a = -r\dot{\theta}^2 e_r$$

$$|a| = \left| -r\dot{\theta}^2 \right|$$

$$= (50)\left(\frac{3}{10}\right)^2$$

$$= \frac{9}{2}\text{ meters/sec}^2$$

The magnitude of the particle's acceleration is therefore $\frac{9}{2}$ meters/sec².

These examples serve to illustrate that an effective problem-solving procedure is independent of any particular engineering course; the same *principles* can be applied to many different areas.

7.2 SOLVING PROBLEMS OF TYPE B: WORD PROBLEMS

As the name suggests, *word problems* are (mathematics) problems posed in English rather than in mathematical language. Most real-world problems are first posed as word problems, since they arise naturally from everyday descriptions of physical situations.

Word problems are generally regarded as more difficult than Type A problems, the reason being that before any mathematics can be applied to the solution of a word problem, the problem itself must first be translated from English into mathematics. This process is called *mathematical modeling* and is by far the most challenging part of the solution of a word problem. There are three stages involved in solving a Type B problem:

Stage 1   Mathematical modeling

Stage 2   Mathematical analysis

Stage 3   Interpretation of the (mathematical) solution in terms of the underlying physical problem

In most introductory engineering courses, Stage 1 is concerned with changing a word problem into a problem of Type A (i.e., translating the English into mathematical language), which is then solved using the application of material learned in the course (Stage 2). Stage 3 recognizes that the (mathematical) solution has some physical significance, since word problems deal mainly with practical, real-world situations. This stage is concerned with interpreting the solution in the context of the original real problem. The following are typical examples of word problems:

EXAMPLE 7.3.   If 1800 cm² of cardboard is available to make a box with a square base and an open top, find the dimensions of the box with largest possible volume.[1]

EXAMPLE 7.4.   An athlete runs around an elliptical track at a constant speed v_0. Find the components of her acceleration at point A in the following figure:

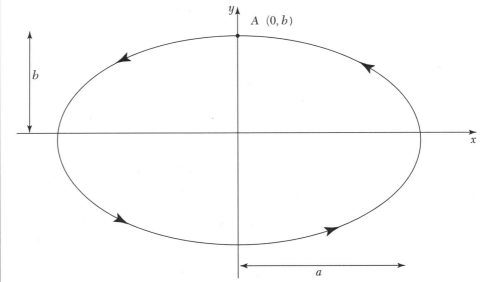

Figure 7.2.   Athlete runs an elliptical track.

EXAMPLE 7.5.   A wooden box filled with rocks is being pulled up a hill by the application of a force $\boldsymbol{F}$ as shown in the accompanying diagram. Find the magnitude of the force $\boldsymbol{F}$ as soon as the box begins to slide.

---

[1]Reprinted with permission from *How to Study Mathematics* by P. Schiavone, Prentice-Hall Canada, 1998

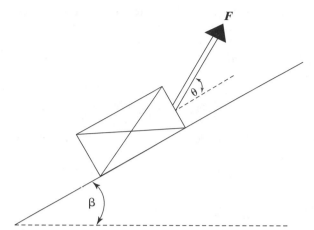

Figure 7.3. Wooden box filled with rocks being pulled up a hill.

Each of these examples hides a corresponding mathematics problem, and the first step in solving any of them is to reveal that problem. By their very nature (mainly because of the mathematical modeling), word problems do not lend themselves to solution by a set procedure. That is, there are no hard-and-fast rules that will guarantee success in solving these problems every time. However, there are a number of general steps that are extremely effective in *Stage 1* of the problem-solving process:

1. Read the problem carefully and at least twice.
2. Try to identify
   i. What you *want*–in other words, the *unknown(s)*.
   ii. What you *have*–in other words, what you *know* or what you are *given* in the problem. In this respect, look for key words. These words will allow you to cut through the padding and reveal the significant information.
3. In most cases, it will be useful to *draw a diagram* to help you *see* what is going on.

   Next comes the difficult part: the *translation* from English into mathematics. Despite the fact that word problems are not *Type A* problems, there *will* be some clues about the underlying theory to be used:

4. Look for *key words* related to known concepts. Consider, for example, the following information regarding Examples 7.3 through 7.5:

EXAMPLE 7.3. | The word "largest" suggests some sort of *maximum- or minimum-value problem* that will involve *differentiation of functions.*

EXAMPLE 7.4. | We are asked to find "acceleration" around an elliptical path. There are known formulas for the acceleration of a *particle* around a *curvilinear path.*

EXAMPLE 7.5. The word "slide" suggests something to do with friction. This problem will probably involve *equations of equilibrium and friction.*

5. Introduce a suitable mathematical notation, preferably similar to that used in class or in the textbook. Choose symbols for the unknown quantities and the given information. For example, use *V* for volume, *a* for acceleration, and so on.

6. Use mathematics to relate *what you want to what you know:* A relation that will connect the given information to the unknown variable and enable you to calculate its meaning. This will be your *mathematical model.*

At the end of this final step, you should have a *Type A* problem. Next comes *Stage 2* of the process. This time, however, you should be on familiar territory. Just follow the procedure for solving *Type A* problems outlined in Section 7.1. *Stage 3* of the process will involve some sort of physical interpretation of your (mathematical) result. In this respect, you should know the meanings of the various mathematical terms used in arriving at the final solution. This will allow you to rule out any (mathematical) solutions that do not make sense (physically) in the context of the situation described in the problem. Such information is usually available from class notes, the textbook, or the problem itself.

There is one significant factor that makes word problem solving *easier* and more effective:

Experience

The more word problems you attempt and solve, the more familiar you will become with the process of mathematical modeling, the identification of key words, and the way in which you link *what you want to what you know* in some relation that will enable you to find the unknown value of the variable. These are all key ingredients of an effective (word-) problem-solving process. Consequently, once again, it comes down to *practice* leading to pattern recognition, fluency, and, finally, effectiveness in problem solving.

SOME OTHER USEFUL IDEAS FOR SOLVING WORD PROBLEMS

- *Similar problems.* As in the case of *Type A* problems, it is often possible to find similar, related, or analogous (solved) word problems in either class notes or the textbook. For example, these problems may have the same key words as the problem you are trying to solve. If so, use the solution to furnish clues on how to proceed in solving your particular word problem (particularly for the mathematical-modeling part).

- *Subproblems.* Sometimes it is necessary to split a more complicated problem into a series of simpler *subproblems,* each with its own *subgoal.* Each subproblem is solved in turn, and the complete solution is built up from each of these *subsolutions.* Problem solving is all about reducing complicated problems to simpler ones. Word problems are no exception!

- *Write it all down!* With word problems, it is even more crucial to record, on paper, each stage of the problem-solving process. (See Section 5.5.) Whenever you figure out how to (mathematically) model a physical situation, you should record exactly the reasoning that led to the model. This is usually the most difficult part of the solution, and you may wish to apply similar reasoning to other word problems in the future. In this regard, the great French mathematician and philosopher Rene Descartes once said,

"Every problem that I solved became a rule which served afterwards to solve other problems."

Also, a clear, logical procedure will allow you to check each stage of the process if an error is found somewhere in the solution.

- *Identify the section of the course material with which the problem is associated.* When all else fails, clues relating to the techniques to be used to model and solve a word problem can be found in some very obvious places. For example, if the problem comes from a textbook, ask yourself the following question: With which section of the textbook is this (or a similar) problem associated? That is, in which particular section does this problem appear? For example, if the problem appears in a section dealing with friction, you can be confident that the problem will be solved using ideas from the theory of friction. If the problem comes not from the textbook, but from class materials (e.g., a handout), note which class material it is intended to reinforce and which particular topic it is intended to cover. These all provide clues to the appropriate theories that should be used to solve word problems.

- *Ask for help.* If you find yourself completely stumped and unable to start a particular word problem, get some help. Try brainstorming with your study group or a study partner. If that doesn't work, seek help from someone with experience in solving this type of problem, for example, a graduate teaching assistant or your professor. Often, identifying the key words or just a simple clue indicating where to look is sufficient to get you started.

- *Don't be fooled by solutions.* Perhaps more than any other type of problem-solving procedure, an effective procedure for solving word problems requires actual *hands-on* experience. It is crucial that you attempt and solve word problems (eventually) yourself. It would be a mistake to read through someone else's solution and conclude that you now know how to do it. The solutions will always look easier than expected, primarily because the most demanding (and significant) component of the solution of word problems is the mathematical modeling. This usually requires considerable effort in thought and concentration, the majority of which is not reflected in the final solution.

- *Evaluate your answers.* Engineering problems are *real* problems. This means that they represent real physical situations, which gives you the added advantage of having even more information with which to check and formulate your answer. In addition, you should remember that engineers do not enjoy the luxury of solving problems just for the sake of solving them; they have to *use the answers* to design, build, and develop actual physical products that will be used in the real world. For these reasons, you should always ask yourself the following questions after you believe you have solved a problem:
  —Is the answer logical?
  —Is the answer reasonable?
  —Is the answer physically possible?
  —What is the (real, physical) meaning of the answer obtained?

It is important that you learn to evaluate your answers before accepting them as true. For example:

—A negative answer is logical if it represents the speed of an object (the object is simply traveling in the direction opposite to the one taken as positive), but, ordinarily, not if it represents time.

—If a cone-shaped drum containing 1 liter of water is pierced on its side, a water loss of 2 liters after 3 seconds is not reasonable.

—It is not physically possible for a rod to expand transversely if it is stretched longitudinally.

In what follows, we illustrate the preceding problem-solving principles by discussing each of Examples 7.3 through 7.5.

EXAMPLE 7.3. If 1800 cm² of cardboard is available to make a box with a square base and an open top, find the dimensions of the box with largest possible volume.

SOLUTION

Consider first, what we want, what we know and try to connect the two with information that enables us to calculate the unknown.

We want to know the dimensions of the box that will lead to the greatest volume.

We know that the dimensions of the box are constrained by the fact that only 1800cm² of material is available to make the box.

We begin by drawing a diagram and using suitable notation to translate what we want and what we know into mathematics. Let b be the length of the side of the (square) base of the box and h the height of the box.

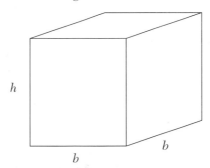

Figure 7.4. Cardboard box problem.

We want b and h so that volume is at a maximum

We know The surface area of the box is to be 1800cm². Noting that the box has a square base and an open top, the surface area is composed of the sum of the areas of the base (b^2) and the four sides, each of which have area hb. Hence, the surface area S of the box is given by

$$S = b^2 + 4hb$$

This is to be 1800cm². Consequently, we can write

$$1800 = b^2 + 4hb \qquad (7\text{-}1)$$

To connect what we want to what we know, we note that the volume of the box is given by

$$V = b^2h \tag{7-2}$$

More precisely, from (7-1),

$$h = \frac{1800 - b^2}{4b}, \, b > 0 \tag{7-3}$$

Using this in (7-2) leads to

$$V = \frac{b^2(1800 - b^2)}{4b}, \, b > 0$$

$$= 450b - \frac{b^3}{4}, \, b > 0 \tag{7-4}$$

This is now our mathematical model which will enable us to find what we want. The first step is to find the value of b which maximizes the expression (7-4) for V. Once this value of b has been found, we use (7-3) to find the corresponding value of h.

We begin by noting that the domain of the function $V(b)$, given by (7-4), is (0, $\sqrt{1800}$] (otherwise, $V < 0$). Since this interval is not closed, we cannot use the "Extreme Value Theorem." However, since V is continuous on its domain, we can use the "First Derivative Test for Absolute Extema". This requires that we first find all critical points of the function V. The set of critical points consists of those values of b obtained from equating $\frac{dV}{db}$ to zero and those values of b at which $\frac{dV}{db}$ does not exist. From (7-4), V and its derivatives are polynomials (and therefore continuous). Hence, there are no values of b at which $\frac{dV}{db}$ does not exist. Consequently, all critical points of the function V arise from equating $\frac{dV}{db}$ to zero:

$$V(b) = 450b - \frac{b^3}{4}$$

$$\frac{dV}{db} = 450 - \frac{3b^2}{4} \tag{7-5}$$

Hence,

$$\frac{dV}{db} = 0$$

$$\Leftrightarrow b = \pm 10\sqrt{6}$$

We eliminate the value $b = -10\sqrt{6}$ since we require that $b > 0$. Alternatively, we can argue that the physical nature of the problem requires that we eliminate the (mathematical) solution $b = -10\sqrt{6}$ since the box cannot have a negative dimension (again, the mathematical model has given us extra information—we select that which is relevant using the information provided in the problem). It follows that V has only one critical point, $b = 10\sqrt{6}$. From (7-5),

$$\frac{dV}{db} > 0, b < 10\sqrt{6}$$

$$\frac{dV}{db} < 0, b > 10\sqrt{6}$$

In other words, V is increasing for all b to the left of the critical point and V is decreasing for all b to the right of the critical point. From the "First Derivative Test for Absolute Extrema", it follows that V is at an absolute maximum when $b = 10\sqrt{6}$. From (7-3), the corresponding value of h is given by

$$h = \frac{1800 - b^2}{4b}$$

$$= \frac{1800 - (10\sqrt{6})^2}{4(10\sqrt{6})}$$

$$= 5\sqrt{6}$$

To summarize, the largest possible volume is given by

$$V_L = V(10\sqrt{6})$$

$$= 3000\sqrt{6}$$

$$\sim 7348.5 \text{ cm}^3$$

occurring when

$$h = 5\sqrt{6} \text{ cm and } b = 10\sqrt{6} \text{ cm}$$

EXAMPLE 7.4. An athlete runs around an elliptical track at a constant speed v_0. Find the components of her acceleration at point A in the following figure:

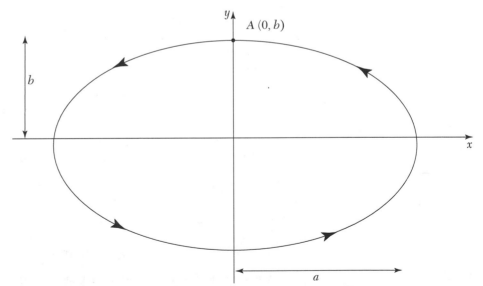

Figure 7.2. *(repeated)* Athlete runs an elliptical track.

SOLUTION
Consider first, what we want and what we know and try to connect the two with information that enables us to calculate the unknown.

   We want: The components of the athlete's acceleration

We know

- Formulas (in different coordinate systems) for the acceleration of a particle moving in a general curvilinear path;
- That the track is elliptical
- That the athlete has a constant speed.
- That since the path is curvilinear, constant speed does not necessarily imply zero acceleration—there will be a non-zero component of acceleration arising from the shape of the track.

We begin by choosing the most appropriate formula for the acceleration of a particle in curvilinear motion.

We choose the normal-tangential representation since it involves the particle's speed and the radius of curvature of the path—one of which is given (the speed) and the other of which can be calculated (the radius of curvature). Hence, in this case, the acceleration of the athlete can be described by

$$a = \dot{v}\,e_t + \frac{v^2}{\rho}\,e_n \tag{7-6}$$

Here e_t, e_n and ρ are, respectively, the tangential direction, normal direction (towards the center of the elliptical path see Figure 7.2) and the radius of curvature of the path. Since the athlete's speed is constant, $\dot{v} = 0$ and the tangential component of the athlete's acceleration is always zero.

Consequently, from (7-6)

$$a = \frac{v^2}{\rho}\,e_n \tag{7-7}$$

Hence, it remains to evaluate only the normal component of the athlete's acceleration at the required point A.

To do this, we need to find the radius of curvature ρ of the elliptical path in a form that can be used in (7-7).

$$\rho = \left| \frac{[1 + (y')^2]^{3/2}}{y''} \right| \tag{7-8}$$

where y is given by the equation of the elliptical path. In fact, the elliptical path can be described by the equation of an ellipse:

$$\frac{x^2}{a^2} + \frac{y^2}{b^2} = 1, a, b > 0$$

or

$$b^2x^2 + a^2y^2 = a^2b^2 \tag{7-9}$$

We need to obtain y' and y''. In fact, differentiating (7-9) implicitly with respect to x, we obtain:

$$y' = -\frac{b^2 x}{a^2 y}$$

$$y'' = -\frac{b^4}{a^2 y^3}$$

Consequently, from (7-8),

$$\rho = \left| \frac{\left[1 + \left(-\frac{b^2 x}{a^2 y}\right)^2\right]^{3/2}}{-\frac{b^4}{a^2 y^3}} \right|$$

At the point of interest, $A(0,b)$, $x = 0$, $y = b$, so that

$$\rho = \left| \frac{\left[1 + \left(-\frac{b^2 x}{a^2 y}\right)^2\right]^{3/2}}{-\frac{b^4}{a^2 y^3}} \right|$$

$$= \left| \frac{a^2 b^3}{b^4} \right|$$

$$= \frac{a^2}{b}$$

Finally, from (7-7),

$$\boldsymbol{a} = \frac{v^2}{\rho} \boldsymbol{e}_n$$

$$= \frac{v_0^2}{\left(\frac{a^2}{b}\right)} \boldsymbol{e}_n$$

$$= \frac{v_0^2 b}{a^2} \boldsymbol{e}_n$$

SUMMARY

The tangential component of the athlete's acceleration is zero. The normal component of the athlete's acceleration is $\frac{v_0^2 b}{a^2}$ (directed towards the center of the elliptical path–centripetal acceleration)

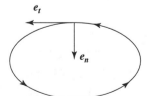

Figure 7.5. Normal component of athlete's acceleration.

EXAMPLE 7.5. A wooden box filled with rocks is being pulled up a hill by the application of a force **F** as shown in the accompanying diagram. Find the magnitude of the force **F** as soon as the box begins to slide.

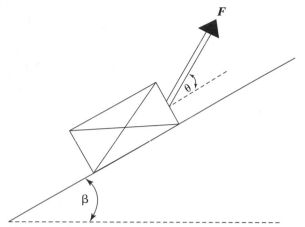

Figure 7.3. *(repeated)* Wooden box filled with rocks being pulled up a hill.

SOLUTION

We want the magnitude F of the force **F**

   We know that the box is being subjected to the force **F** and is just beginning to slide.

   We must use existing theory to relate what we **want** to what we **know**. In this case, we utilize the force equations of equilibrium for the box. To get a picture of the different forces acting, and to make sure that we don't miss anything, we first draw a free body diagram:

- **Draw a free body diagram**

   Figure 7.6. Free body diagram of Figure 7.3.

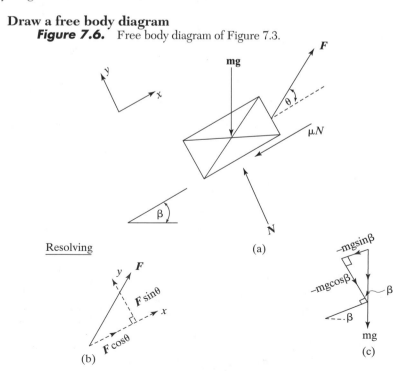

In the free body diagram in Figure 7.6, **mg** represents the weight of the box, **N** the normal reaction, μ the coefficient of friction and μN the magnitude of the frictional force as the box begins to slide.

- **Resolve forces in two orthogonal directions:** up the slope (direction of motion) and orthogonal to the slope

In what follows,

$$F = |\boldsymbol{F}|, N = |\boldsymbol{N}|, mg = |\boldsymbol{mg}|$$

First resolve up the slope in the positive x direction. From the free body diagram (Figure 7.6):

$$\nearrow + : F\cos\theta - mg\sin\beta - \mu N = 0 \qquad (7\text{-}10)$$

Next, resolve in the orthogonal y direction. Again, from the free body diagram:

$$\nwarrow + : F\sin\theta - mg\cos\beta + N = 0 \qquad (7\text{-}11)$$

- **Solve for the magnitude F of the force F**

From (7-11)

$$N = mg\cos\beta - F\sin\theta \qquad (7\text{-}12)$$

Using (7-12) in (7-10), we obtain

$$F\cos\theta - mg\sin\beta - \mu(mg\cos\beta - F\sin\theta) = 0$$
$$F(\cos\theta + \mu\sin\theta) = mg(\sin\beta + \mu\cos\beta)$$

Finally,

$$F = \frac{mg(\sin\beta + \mu\cos\beta)}{(\cos\theta + \mu\sin\theta)}$$

- **Interpretation**

As expected, the magnitude of the force **F** required to make the box slide depends on the incline of the slope, the coefficient of friction μ, the weight of the box and the angle at which the force **F** is applied.

Problems

1. Look at the last assignment from each of your engineering courses. Identify each problem as being of either *Type A* or *Type B*, as defined in this chapter. To which category do the more difficult problems belong? Which courses tend to ask more *Type A* problems? Which courses tend to ask more *Type B* problems? Why?

2. Describe your own approach to solving word problems(*Type B* problems). How does it differ from that outlined in Section 7.2?

3. Where do you encounter most difficulties in solving word problems? Explain. Develop a strategy for dealing with these difficulties. Implement that strategy when you solve a new word problem.

4. In view of what has been said in this chapter, what would you change about your own approach to solving word problems?

5. Find a word problem similar to those in Examples 7.3 through 7.5. Use the procedure in Section 7.2 to solve the problem. Make sure you record each step of the solution procedure.

6. Consider the two sub-problems in Problem 13 of Chapter 4. Solve each by using the step-by-step procedure outlined in Section 7.2.

7. Make up a word problem of your own. Make sure to provide a detailed solution to your problem. Working backwards, show how the solution can be developed using each of the steps outlined in Section 7.2.

8. Write a 500-word paper about "Word Problems: Why They Arise and How to Solve Them."

9. Write a one-page paper on "The role of the *Type A* problem in undergraduate engineering."

10. On your next assignment, identify each problem as being of *Type A* or *Type B*. Implement the appropriate solution procedure and solve each problem using the step-by-step analysis mentioned in this chapter.

11. Find *real-life* examples that lead to problems of *Type A*. Do the same for *Type B*.

12. Why do word problems almost always arise naturally in engineering? Explain.

13. What do you think are the 10 most important skills required for effective problem solving? Make a list. Rank these skills from 1 (most important) to 10 (least important). Which skills are your weakest? Which are your strongest?

14. Do you solve problems more effectively on your own or in collaboration with others? Which approach do you think is more common in industry? Why?

8

Mathematics

Mathematics is the *language* of engineering. It is the vehicle by which ideas are analyzed, developed, and communicated in engineering. We have already seen this in Chapter 7, where we used mathematical modeling to solve word problems. It is no accident, therefore, that the undergraduate engineering curriculum includes several mathematics courses, each one designed to give you adequate skill and knowledge to deal with engineering problems of increasing complexity. (See Figure 8.1.)

Besides furnishing you with the tools of the profession, engineering mathematics courses are intended to develop the logical, rational, problem-solving skills that are so crucial to good engineering practice. Some of the best mathematicians I know are engineers. For example, my colleague David Steigmann is a professor of mechanical engineering at the University of California at Berkeley. As a research engineer, David is interested in mathematics primarily as a *means to an end;* his main interests lie in solving problems related to continuum mechanics, shell theory, elasticity, the stability of mechanical structures, surface stress in solids, capillary phenomena, and the mechanics of thin films. Yet, in analyzing these problems, he has developed significant mathematical expertise, which puts him among the best in the world!

SECTIONS

- 8.1 How to Succeed in Mathematics Courses

OBJECTIVES

In this chapter you will:

- Learn about the importance of mathematics in engineering.
- Find out how to make sure that you are well-prepared for all engineering mathematics courses.
- Discover techniques for maximizing your performance in calculus.
- Learn how to use mathematics to solve engineering problems.

You should regard your mathematics courses in the same way—as a *means to an end*—courses designed to equip you with the necessary skills to succeed in engineering. You don't have to love mathematics or even appreciate its beauty. Just learn how to use it effectively to solve engineering problems—that's the key.

8.1 HOW TO SUCCEED IN MATHEMATICS COURSES

The study strategies discussed so far in this book are applicable to *all* your engineering courses, including mathematics courses. However, there is one particular area that merits extra attention when it comes to mathematics courses, namely, *preparation:*

> *Mathematics, to a greater extent than most engineering disciplines, is a* **cumulative** *subject: The understanding of one part depends heavily on the understanding of previous parts.*

As a mathematics instructor, I found that the single most common reason that students do not succeed in mathematics courses is a lack of adequate preparation. The cumulative nature of the subject makes it absolutely essential that you take the necessary steps (Section 5.2) to ensure that you are fluent in the basic grammar and vocabulary of the language. To appreciate why, consider the following simple analogy. You pay $100 to enroll in a special seminar called "How to Make a Million Dollars." You are informed the day before the seminar that, due to a mix-up, you have been allocated a seat in the French version of the seminar, and there are no seats left in the English version. You are faced with a choice: You can either cancel your registration and get the $100 back or go to the French version of the seminar. Attending the seminar is a unique opportunity, as the speaker will not return for another year and you know people who have made a million dollars from the knowledge they gained from the seminar, so you really want to attend. You previously studied French and were able to speak the language quite fluently. In fact, you were in France only three months ago, when you became quite comfortable with the language. However, since returning home, you have had no reason to speak French, as all of your friends and family speak only English. Still, believing that the fluency you held three months ago will get you through, you elect to attend the French version of the seminar. Unfortunately, when you get there, you discover that the speaker speaks so fast that you experience great difficulty keeping up. You manage to understand the odd word here and there, but your French is worse than you expected. Things deteriorate further as the seminar progresses: Charts, handouts, and the interaction among the people are just too much for your limited French skills. You leave the seminar having learned nothing and having wasted $100.

If we identify the main problem in this scenario, we quickly see that it has little to do with your proven competence in French (after all, you spoke the language fluently only a few months ago). Rather, it has more to do with the fact that you are trying to perform *today* in a language in which you were fluent *three months ago*. With the basics no longer at your fingertips, how could you expect to *use* the language to learn anything?

Unfortunately, the same scenario is played out year after year with engineering students entering mathematics courses at all levels, particularly first-year courses. These students expect to perform at a high level of sophistication in a language that they have mostly forgotten. This wouldn't be so bad if the professor incorporated time for significant review into the course. However, constraints on resources almost always make this impossible. Thus, professors will assume that the responsibility for fluency in the specified prerequisite mathematics rests *solely with you,* and the professor will carry on with the new material regardless. The ill-prepared students usually realize they are in trouble around midterm time when exam grades begin to tumble. By then, it is almost always too late to make up the lost time (and grade).

To avoid this happening to you, take the advice given in Section 5.2, and ensure that you are equipped to learn new mathematics from the first day of classes. This does not take long. Only an hour or so practicing the necessary skills will give incredible benefits and significantly affect your performance.

Figure 8.1 shows a typical sequence of mathematics courses in an undergraduate engineering curriculum. The figure is arranged in the shape of a pyramid to demonstrate the cumulative nature of the mathematics sequence.

Notice that the base or foundation of the mathematics sequence (the pyramid) is preuniversity mathematics—that is, the basic skills we learn in high school. If this foundation is weak, any subsequent additions to the pyramid will make the whole thing unstable. So it is with mathematics: As mentioned earlier,

> The understanding of one part depends heavily on the understanding of previous parts.

The following procedure is based on material from previous chapters and should be applied to all mathematics courses at any level.

PROCEDURE FOR BEING EFFECTIVE IN MATHEMATICS COURSES

1. Identify and become fluent in required or assumed prerequisite skills. (See Section 5.2.)
2. Maintain effectiveness throughout the course by:
    *   Making the most of class time (lectures, labs, etc; see Section 3.1.)
    *   Making effective use of the textbook. (See Section 5.4.)
    *   Being effective on assignments. (See Section 5.5.)
    *   Getting help. (See Sections 3.2 and 5.7.)
    *   Following effective problem-solving procedures. (See Chapter 6.)

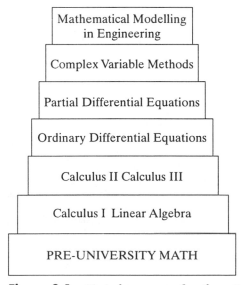

Figure 8.1. Typical sequence of mathematics courses.

3. Review for, and take, examinations effectively. (See Chapter 7.) In particular, we mention that in beginning mathematics courses (e.g., calculus, linear algebra, ordinary differential equations), examinations are largely predictable, mainly because of the limited range and repetitive nature of the topics covered. As a result, the same types of questions appear over and over again. (See, for example, the following sample final exams.) Once the most significant areas of expertise have been identified, the task reduces to working through as many relevant problems as possible, that is, problems similar to those likely to be asked on the actual examinations—preferably, practice examination problems. In working through these kinds of problems you will become fluent in the relevant techniques and begin to see patterns in the solutions. Eventually, the solution process becomes so systematic, that the actual examination becomes an anticlimax. There are two key ingredients in this procedure:

- A good supply of practice examination problems.
- Full, easy-to-read solutions to every problem.

These two ingredients make all the difference! It is not difficult to obtain old examinations, but fully worked-out, comprehensive solutions are almost never available. And even when they are, they are usually not well presented and are often confusing.

CALCULUS I[1]
FINAL EXAMINATION #1

Time: 2 hours $\hspace{4cm}$ **Level of difficulty:** 4/5

Value

(10%)

1. (i) Evaluate the following limits.

   (a) $\lim\limits_{x \to \infty} (\sqrt{x^2 + 3x - 1} - x)$

   (b) $\lim\limits_{x \to -\infty} (\sqrt{x^2 + 3x - 1} - x)$

   (c) $\lim\limits_{h \to 0} \dfrac{\sin \pi h^2 \tan 3h^2}{h^3}$

   (ii) Let

$$f(x) = \begin{cases} 1 - \sin x, & -2\pi \leq x < 0, \\ 1, & x = 0, \\ x^2 - 1, & 0 < x \leq 2. \end{cases}$$

   (a) Is $f(x)$ continuous at $x = 0$? Explain.

   (b) Is $f(x)$ differentiable at $x = 0$? Explain.

(15%)

2. Evaluate the following definite integrals:

   (a) $\displaystyle\int_0^1 |x^3 - x - 3x^2 + 3|\, dx$

   (b) $\displaystyle\int_0^1 \dfrac{2x^3 + 9x^2 + 5}{(x^4 + 6x^3 + 10x + 7)^{1/2}}\, dx$

(c) $\int_{-1}^{1} x^{-8/9} \sin(x^{1/9})\, dx$

(d) $\int_{-a}^{a} x^{3} \cos(x^{3})\, dx, \quad a \in IR.$

(15%) 3. Find the following.

(a) $\int \dfrac{\cos^{3}x \sin x}{\sqrt{1 + \cos^{4}x}}\, dx$

(b) $\int \dfrac{x^{-3} + x^{6}\sqrt{1 + x^{5}}}{5x^{2}}\, dx$

(c) $\int \dfrac{\cos\left(\dfrac{1}{t}\right)}{t^{2}}\, dt$

(d) $\int \sin^{4}t \cos^{3}t\, dt$

(e) $f'(x)$ where $f(x) = \displaystyle\int_{x}^{x^{2}+1} \tan(\sqrt{u^{2} + 1})\, du$

(20%) 4. Discuss the graph of $y = f(x) = \dfrac{7x^{2}}{x^{2} - 1}$ under the following headings.

(a) Domain of $f(x)$.

(b) x and y-axis intercepts.

(c) Asymptotes.

(d) Intervals of increase or decrease.

(e) Local maximum and minimum values.

(f) Concavity and points of infection.

Sketch the curve.

(10%) 5. An open-top cylindrical can is made to hold b liters of oil. Find the dimensions that will minimize the cost of the metal.

(10%) 6. Find the limit if it exists. If the limit does not exist, explain why.

(a) $\displaystyle\lim_{\theta \to 0} (\theta^{2} \cot \theta + 5)$

(b) $\displaystyle\lim_{x \to 0} \left(\dfrac{1 - \sin x}{x^{3}} \right)$

(c) $\displaystyle\lim_{x \to 1^{+}} \dfrac{1 - \sqrt{x}}{1 - x}$

(d) $\displaystyle\lim_{x \to -\infty} \dfrac{x - 1}{x^{2}}$

(10%) 7. Find the area bounded by the graphs of $y = x^{3}$ and $y = x$.

(10%) 8. The height h of a triangle is increasing at a rate of 5 cm/min while the area of the triangle is increasing at a rate of 8 cm² / min. At what rate is the base of the triangle changing when the height is 50 cm and the area 500 cm²?

CALCULUS I²
FINAL EXAMINATION #2

Time: 2 hours

Level of difficulty: 4/5

Value

(30%)

1. Evaluate when possible:

   (a) $\lim\limits_{x \to 0} \dfrac{x - x\cos 3x}{x + x\cos 3x}$ **OR*** $\lim\limits_{x \to 0} \dfrac{\sin x - x - x^3}{x^2}$

   (b) $\lim\limits_{x \to 2} \left(\dfrac{1}{x - 2} - \dfrac{2}{x^2 - 4} \right)$ **OR*** $\lim\limits_{x \to 0} \dfrac{(\ln x)^2}{x}$

   (c) $\lim\limits_{x \to -\infty} \dfrac{8x^3 + 2x^2 - 1}{(x^9 - 4x + 11)^{1/3}}$ **OR*** $\lim\limits_{x \to 0^+} (3x + 1)^{2/x}$

   (d) $\displaystyle\int_0^2 |x^3 - 3x^2 + 2x|\, dx$ **OR*** $\displaystyle\int_1^8 e^{\sqrt[3]{x}} x^{-2/3}\, dx$

   (e) $\lim\limits_{n \to \infty} \displaystyle\sum_{i=1}^{n} \dfrac{8}{n}\left[3\left(2 + \dfrac{2i}{n}\right)^2 + 2\left(2 + \dfrac{2i}{n}\right)\right]$.

(20%)

2. Find: (a) $f'(x)$ if $f(x) = [10 + (1 + \sqrt{3 + x})^{3/2}]^{3/2}$

   OR* $f(x) = e^{\sqrt{x}} \cos(x^2)$

   (b) $g'(x)$ if $g(x) = \cos[\tan(\sin(x^2 + 1))]^{1/2}$

   OR* $g(x) = \ln[\ln(x^4)]$

   (c) $\dfrac{dy}{dx}$ if $x^2y = \cos(xy)$

   OR* $e^{x\sin y} + x^3y^2 = \ln(x^2 + y^2)$

   (d) $y'(x)$ if $y(x) = \displaystyle\int_{\sqrt{x}}^{1} \sqrt{2 + \tan^2(t^2)}\, dt$

   (Do not simplify your answers.)

(25%)

3. Find: (a) $\displaystyle\int x^3 \sin(3x^4 + 10)\, dx$ **OR*** $\displaystyle\int_{e^2}^{e^3} \dfrac{dx}{x \ln x}$

   (b) $\displaystyle\int \sec^2 x \tan^3 x\, dx$ **OR*** $\displaystyle\int_0^{\pi/2} \dfrac{\cos x}{1 + \sin x}$

   (c) $\displaystyle\int_{-\pi}^{\pi} \dfrac{t^4 \sin t}{1 + t^8}\, dt$ **OR*** $\displaystyle\int_0^1 \dfrac{e^{\sin^{-1} x}}{\sqrt{1 - x^2}}\, dx$

   (d) $\displaystyle\int \dfrac{\sqrt{1 + x^{-5}}}{x^6}\, dx$ **OR*** $\displaystyle\int \dfrac{e^x}{e^x + 2}\, dx$

   (e) $\displaystyle\int \sin x[\sin(\cos x)]\, dx$ **OR*** $\displaystyle\int \dfrac{\sqrt{\arctan x}}{1 + x^2}\, dx$

(10%)

4. Sketch the graph of $y = f(x) = \dfrac{x}{x - 1}$ **OR***

   $y = f(x) = \ln(\cos x)$ identifying any critical points, asymptotes, x and y-intercepts, points of inflection and local maxima and minima.

*For students taking early transcendentals—although the transcendentals options are important, you should practice with **both** versions of this exam.

(8%)   5. Find the total area bounded by the curves $y = x - 2$ and $y^2 - 2x - 4 = 0$.

OR°

Find the Taylor polynomial of degree three, $T_3(x)$, for the function $f(x) = e^x \sin x$ about the point $c = 0$.

(7%)   6. Find the ratio of the height (h) to the radius (r) of the minimum surface-area cone of constant volume b.

(Hint: Surface area S of cone is given by $S^2 = \pi^2(r^4 + r^2h^2)$ and $b = \dfrac{1}{3}\pi r^2 h$.)

INTRODUCTORY ORDINARY DIFFERENTIAL EQUATION[3]
FINAL EXAMINATION #1

Time: 2 hours   **Level of difficulty:** 3/5

1. (20%) Solve the following differential equation

$$x^2\frac{d^2y}{dx^2} + x\frac{dy}{dx} - y = x^2 \ln x, \quad x > 0$$

2. (10%) Use the method of *reduction of order* to find the general solution of the differential equation

$$\frac{d^2y}{dx^2} - 4\frac{dy}{dx} + 4y = \frac{e^{2x}}{x^2}, \quad x > 0$$

3. (10%) Determine the appropriate form for a particular solution of the differential equation

$$(D - 1)^4(D^2 + 16)^2y = xe^x + x\cos 4x$$

when using the method of undetermined coefficients. **Do not evaluate the coefficients.** Note that, here, $D^n \equiv \dfrac{d^n}{dx^n}$.

4. Find the Laplace transform of each of the following functions.
    (i) (10%)

$$g(t) = \begin{cases} \cos t, & 0 < t < \pi, \\ 0, & \pi < t < 2\pi, \end{cases}$$

$g(t) = g(t + 2\pi), t \geq 0$.
(ii) (5%)

$$h(t) = e^t \cos 2t + t \sin t$$

5. (15%) Solve the following initial value problem using the method of Laplace transforms.

[1,2] Reprinted with permission from: "Calculus Solutions—How to Succeed in Calculus—From Essential Prerequisites to Practice Examinations" by P. Schiavone, Prentice-Hall, Canada, 1997

$$\frac{d^2x}{dt^2} + x = \begin{cases} 2, & 0 \le t < \frac{\pi}{2}, \\ 0, & t \ge \frac{\pi}{2}, \end{cases}$$

$$x(0) = 0, \frac{dx}{dt}(0) = 2$$

6. (10%) Use the convolution theorem and the method of Laplace transforms to solve

$$\frac{d^2y}{dx^2} + 6\frac{dy}{dx} + 9y = H(x), \quad y(0) = 0, y'(0) = 0$$

where $H(x)$ is a known function of x.

7. (20%) Find the general solution in a series about $x = 0$. Give a region of validity for your solution (you must justify your conclusion).

$$(4 - x^2)y'' - 2xy' + 2y = 0$$

INTRODUCTORY ORDINARY DIFFERENTIAL EQUATIONS[4]
FINAL EXAMINATION #2

Time: 2 hours   **Level of difficulty:** 3/5

1. (10%) Solve the following nonlinear ordinary differential equation.

$$y\frac{d^2y}{dx^2} - \left(\frac{dy}{dx}\right)^2 = 0, \; y > 0$$

2. (15%) For the following homogeneous differential equation find a solution of the form x^n, where n is a positive integer.

$$(x^2 - 1)\frac{d^2y}{dx^2} - 2x\frac{dy}{dx} + 2y = 0$$

Use this solution to find the general solution of the following corresponding inhomogeneous equation

$$(x^2 - 1)\frac{d^2y}{dx^2} - 2x\frac{dy}{dx} + 2y = (x^2 - 1)^2$$

3.

(i) (10%) Find

$$L^{-1}\left[\frac{5s + 3}{s^2 + 4s + 5}\right]$$

(ii) (10%) Determine

$$L^{-1}\left[\frac{s}{(s+3)^5(s^2+16)}\right]$$

in the form of an integral (Do not evaluate the integral).

4.

(a) (10%) Find the general solution of the equation

$$y'' - 6y' + 13y = 15\cos 2x \qquad \text{(F2.1)}$$

(b) (5%) If the left-hand side of (F2.1) is replaced by $y'' + 4y$, what would be the corresponding form of particular solution y_p if the method of undetermined coefficients is used to find the general solution?

5. (15%) Solve the following initial value problem using the method of Laplace transforms.

$$\frac{d^2y}{dx^2} + 4\frac{dy}{dx} + 4y = e^{-2x}, \quad y(0) = 0, y'(0) = 1$$

6. (25%) Find the general solution of the following differential equation in terms of series centered at $x = 0$.

$$2x^2(1-x)\frac{d^2y}{dx^2} - x(1+x)\frac{dy}{dx} + (1+x)y = 0, \quad x > 0$$

Problems

1. Write about mathematics and its place in engineering in a one-page paper entitled "Why mathematics is such an essential component of an engineering education."

2. Devise a pyramid similar to that in Figure 8.1 for the mathematics courses required for your particular engineering degree.

3. List five well-known mathematicians who contributed to mathematics mainly as a result of their interest in solving real, physical problems.

4. Talk to the corresponding professor, and identify the most important skills that are a prerequisite for your next mathematics course.

5. Take the assessment test provided on p. 70 (Chapter 5). How fluent are your basic skills in precalculus? From the information given in Problem 4, make up a similar test for your next mathematics course. Take the test just before you take the course.

6. Look at the pyramid of Problem 2. For each mathematics course, describe the relevance of the material presented to your engineering education. Where and why would you use such material?

[3,4] Reprinted with permission from: "Introductory Ordinary Differential Equations" by P. Schiavone, Prentice-Hall Canada, 1998.

9

Developing Engineering Skills

Excellent technical engineering skills are, in themselves, not sufficient to guarantee a successful career in engineering. The following is a list of skills that employers are constantly demanding.

- The ability to communicate effectively, including
  - —The ability to write clear, coherent technical documents.
  - —The ability to present well-researched, well-organized seminars confidently and competently.
- The ability to work effectively in teams with people from different backgrounds.
- The ability to demonstrate creativity in all aspects of the profession.
- The ability to manage personnel and resources in engineering projects.
- The ability to keep up to date with developments during one's years as a professional engineer.

These skills are highly prized by employers. Acquiring such skills will not only give you the edge in employability, but also allow you to go well beyond your training as an engineer towards an ever-expanding number of exciting and challenging opportunities.

OBJECTIVES

In this chapter you will learn:

- The importance of good communication skills in engineering.
- How to improve your communication skills—as in:
  - —the ability to *write* clear, coherent technical documents
  - —the ability to *present* well-researched, well-organized information—confidently and competently.
- What employers look for in engineering graduates.
- About creativity in engineering and how to develop your own creativity.

What can you do to give yourself a head start and acquire the foregoing skills before you graduate? We have already discussed how *teamwork* and *independent learning skills* (Chapter 4) can contribute to success in engineering study. Practicing these skills as an undergraduate engineering student is an excellent way to equip yourself beforehand with at least three of them:

- *The ability to work effectively in teams with people from different backgrounds.* Your experience with teamwork as an undergraduate engineering student will make this skill second nature to you by the time you graduate. Not only will you know how to work as part of a team, but you will have the ability to organize and manage teams.

- *The ability to keep up to date with developments in engineering.* As an engineering student, you become well accustomed to learning and thinking independently. The ability to keep up to date with developments in engineering requires that you find and assimilate information independently, as required. This is exactly what you do as a resourceful, committed undergraduate engineering student.

- *The ability to manage personnel and resources in engineering projects.* Practicing teamwork and independent learning together means that you learn to manage not only your own resources (time, energy, money, etc.), but also the resources of those around you. You learn how to find information effectively, how to use that information to attain a defined goal, and how to communicate the information to other group members.

In this chapter, we concern ourselves with the remaining two engineering skills:

1. The ability to communicate effectively, including
   i. The ability to *write* clear, coherent technical documents.
   ii. The ability to *present* well-researched, well-organized seminars confidently and competently.
2. The ability to demonstrate creativity in all aspects of the profession.

9.1 COMMUNICATION SKILLS

Suppose, after you graduate, you come up with a new idea—an earth-shattering idea for a new product. You spend perhaps six months developing the theory, researching the consequences, and experimenting with the product until all your suspicions come true, and the product works! What do you do next? Do you need to build a bigger lab? Do you need to develop the experiments further? Do you need to secure more funding? Do you need to begin production or convince an industrial tycoon to manufacture your product? Whichever it might be, one thing is sure:

> You will need to convince someone else, perhaps someone with little technical knowledge, that your idea is worthy of his or her time and money.

Basically, an idea remains just an idea until its essence and its implications are communicated to others. This means that, whether you work for yourself or for a major corporation, you need to be able to *sell* your genius, inspiration, enthusiasm, excitement, knowledge, and abilities to others, often people with time and money to invest, but with little patience and understanding for the technical beauty of the product.

Communicating is all about *conveying information* concisely and effectively. As such, effective communication is also a vital aspect of acquiring understanding. (Trying to communicate your knowledge effectively in writing or discussion is an excellent way of reinforcing learning; any teacher will tell you that *teaching is the best way of learning anything!*) Fortunately, you do not need to experiment with effective communication: People have been doing it for years. All you need to do is note what successful communicators do and do the same thing yourself. However, there is one essential ingredient that you cannot avoid: *practice.* Excellent communication skills, like most things, come from practice. Just remember that

> Practice results in improvement!

The two main methods of communication in engineering are *written and oral communication.* As an undergraduate engineering student, you have many excellent opportunities to practice both. For example, you *write* the solutions to problems in lecture notes, term papers, reports, essays, assignments and tests, and so on. Similarly, your oral communication skills are used in oral presentations as part of a team project debates or student seminars interpersonal communications with fellow students and professors, and in engineering student organizations, recreational programs, and student government. Whatever your experience of communicating information is so far, one thing is clear:

> Today's employers demand effective written and oral communication skills from all engineering graduates.

In other words, there is no choice in the matter: Excellent communication skills are *necessary* for success in engineering.

Your time in college is an excellent opportunity to work on and improve your communication skills. As previously mentioned, as an undergraduate engineering student, you are given many different opportunities to practice your communication skills, both written and oral. Take every one of these opportunities and use them to your full advantage to improve your abilities in both aspects of communication. The advantages will be not only immediate (the communication of well-organized, logical, coherent, and well-presented information; see Sections 5.3 and 5.5), but, more importantly, long term, for when you enter the engineering work world. If you don't take advantages of such opportunities now, you will bring your weaknesses in communication into your career, when you will have less time and opportunity for improvement and where you will be competing daily with people who have already mastered the necessary skills.

> The price for waiting until after graduation to learn effective communication skills is much higher!

In this section, we will suggest different ways for you to develop your skills in each area of communication, written and oral.

DEVELOPING EFFECTIVE WRITING SKILLS

Effective writing skills develop slowly, over a period of time, with *practice*. That's why it's so important to begin the process as soon as possible. The following are a few suggestions for developing such skills:

- *Take Writing Classes.* Almost every university or college in the country offers some form of supplementary noncredit (voluntary) writing classes. These classes are an excellent way for you to become more proficient in your writing. They are short, targeted, and to the point, focusing on the mechanics of writing rather than on an appreciation of the English language. In my own university, many short courses (between 4 and 18 hours per semester are offered). The following list gives the names and descriptions of some of those courses:

  —*An introduction to university essay writing.* How to develop a clear prose style; editing for common sentence-level and grammar problems (6 hours)

  —*Editing for correctness and clarity.* An overview of essay writing, focusing on structure, coherence, and basic research skills (4 hours)

  —*Writing for the university.* Covers all aspects of essay writing (18 hours)

  —*Revising and editing a paper.* Designed to help you write clear, coherent, error-free prose (10 hours)

  In addition, you can obtain help on an individual basis through private consultations. These services are usually provided through student services or a related campus organization. I strongly urge you to seek out such classes on your particular campus and take as many as you can—not necessarily all at once, but certainly many over the course of your degree.

- *Write as much as you can.* Apart from formal course work (e.g., assignments, tests, etc.), you can give yourself the following opportunities to practice your writing:

  —*Take lecture notes.* Regard each set of lecture notes (see Section 5.3) as a writing assignment. Rewrite the notes as necessary until they are clear and concise. Have someone read them over and give you suggestions about how they can be improved. In particular, learn to incorporate mathematics into text using established procedures, as is done in technical books and journals.

  —*Write Letters.* Write letters to friends and family back home, to newspapers, to magazines, and so on.

  —*Write Summaries.* Summaries (of class notes, assignments, important theories, etc.) are extremely effective when you are reviewing, but they also provide an opportunity for you to practice your writing skills.

- *Read as much as possible.* Read newspapers, journals, novels, technical books, and magazines. Read at least a few pieces of work every day. There is no doubt that reading contributes to good writing skills, if only for the following reasons:

  —*Reading exposes you to professional writing.* This kind of writing serves as an example for your own writing and exposes you to techniques and procedures required for effective writing.

  —*Reading expands your vocabulary and grammar.* As you encounter new words, look up their meanings and use those same words yourself in conversation and in your writings.

  —*Reading increases your mental agility.* Reading exercises your mind.

- *Look for examples.* We have mentioned many times in this book how it is almost never necessary to "reinvent the wheel." The same is true in the case of writing a good report or proposal. Look at what other people have written. Go to the library, speak to people with experience, or surf the Web and seek out pieces of writing similar to the one you intend to write. Note the structure, use of language, headings, grammar, scientific notation, use of mathematics, how the mathematics is positioned with respect to text, and so on. If you know that a particular report or proposal was successful, then what you have in front of you is a winning combination. Use this information to help you prepare your own work. This procedure is commonplace even among university professors who apply for funding. Usually the funding agency will issue what it refers to as "successful former applications" so that prospective applicants can see how to organize and present their proposals in the most effective manner.

- *Get lots of feedback; have other people read your work.* When you are faced with any writing assignment, get it done early, well ahead of time. That way, you can ask someone (e.g., your professor, a senior student, or a writing instructor) to look it over and provide some helpful suggestions. Then take the time to rewrite the assignment, incorporating all the suggestions made by the reader(s). Remember, the effectiveness of your writing will be judged by how well someone else understands what you write. Consequently, whenever possible, take the opportunity to have someone else read and criticize your work. This serves the same purpose as a rehearsal.

- *Read your own work out loud to yourself.* Reading your own work out loud to yourself will allow you to *hear* what someone else will hear when you present your work. In this way, you can dramatically improve your work, by yourself

- *Elect to take courses that involve a strong writing component.* Whenever you are given a choice of which course's to study (elective's), choose courses that involve extensive writing. They need not be engineering courses. For example, you might choose economics, history, sociology, or a language course. Not only will you have the opportunity to write, but you will have available the skills of a trained professional to provide essential feedback on your writing.

- *Learn word processing.* The presentation of handwritten reports in a professional context is now uncommon. It is therefore essential that you become familiar with the use of a word processor. Popular packages now come with excellent, simple, on-line tutorials, making them extremely easy to learn. Add to that the many different tools available (e.g., automatic spell checking, an on-line thesaurus and dictionary, grammar checking, on-line editing, etc.) and you have an easy way to produce high-quality documents. Word processors also promote the effective use of layout: You have the ability to mimic styles already familiar to you from textbooks and technical journals. Since corrections are so much easier to achieve, word processing will also allow you to concentrate on your ideas and understanding of the material to be presented.

Effective writing is a complex skill and is developed over time. Following the foregoing suggestions and practicing whenever you can will make you not only more competent, but also more confident in your work and in the way you present it.

DEVELOPING EFFECTIVE ORAL COMMUNICATION SKILLS

One of the most rewarding aspects of my profession is being able to follow the careers of my former students as they enter professional engineering in the *real world*. Each time one of them comes to visit me, I am amazed by just how much they have changed. They appear immaculately dressed, polished, confident, and extremely well-spoken. In addition, they are much more comfortable than they have ever been in one-on-one conversation—even those who used to sit at the back of the class and never utter a single word! I often remark, "How times have changed" from the days when they had to be constantly prompted for information and would often feel intimidated just being in the same room as me!

Most of them explain that their metamorphosis was more out of necessity than by choice: Within only a few weeks of beginning their careers, they would find themselves in formal meetings with clients, having lunch or dinner meetings with consultants, participating in conference calls, or explaining a new product line to an audience as part of a formal presentation. In other words, their interpersonal communication skills were now the deciding factor between success and failure. Being engineers, my former students adapted as necessary, but not without a certain amount of anguish, frustration, and embarrassment. Consequently, their message to me is always the same:

> "I wish I had paid more attention to my communication skills while at the university."

There are basically two different types of oral communication in which we engage interpersonal communication and formal presentations.

INTERPERSONAL COMMUNICATIONS

Interpersonal communication includes the following:

- One-on-one informal conversations with a second party
- One-on-one formal meetings
- Group meetings in which you act as a participant or team leader
- Interviews in which you are the *interviewer* or the *interviewee*

Effective interpersonal communication involves many different skills, including:

- listening skills
- the ability to define what you need from a meeting and how to ask for it
- the power of persuasion
- sensitivity to the beliefs and perceptions of others
- the ability to understand and appreciate another's point of view
- the ability to present what you have to say in such a way that it appears attractive to the other party

These and many other aspects of interpersonal communication are discussed in the many different books dealing with personal empowerment and the psychology of success (See, for example, [6]–[9].) These books make an excellent read and are full of extremely useful tips and suggestions for achieving your goals. They also serve as excellent motivators as you encounter the usual obstacles that are commonplace on the path to success in any discipline. Read these or related materials, and use the opportunities provided by the university environment to practice your skills starting today. For example:

- *Take a psychology or communications course.* Learn as much as possible about people, human relations, and how people communicate. This will add to your understanding of interpersonal relationships.
- *Get involved in engineering student organizations.* They offer excellent opportunities to discuss ideas and get talking.
- *Serve on university committees as a student representative.* These committees exist at all levels of university administration and are an excellent way to practice your interpersonal communication in a formal setting.

FORMAL PRESENTATIONS

I have been giving formal presentations now for almost 15 years. To this day, as I am about to "go on stage," I still get nervous, my knees begin to buckle, and my mind goes blank. Despite all of this, I continue to give extremely effective presentations, at all levels, from teaching freshman engineering classes to giving complex research seminars at international conferences.

The fear of public speaking is perhaps the greatest fear of all. The symptoms I experience just before giving a presentation are not uncommon. To quote Mark Twain:

> *"There are two types of speakers: those that are nervous and those that are liars."*

Everyone—even experienced speakers— has some anxiety when speaking in front of a group of people. This is perfectly normal. Accept this simple fact, and then take the necessary steps to make sure you give a superb presentation anyway. The key to doing this lies in

> Preparation!

The following is the first of a list of simple suggestions I offer from my (many) experiences giving (extremely effective and not so effective) formal presentations in industry and academia:

1. *Prepare the technical aspects of your presentation properly and thoroughly well beforehand.* Prepare all subsidiary materials, for example, slides, videos, handouts, and so on, meticulously. Make sure that:
   i. They are technically correct and contain no errors.
   ii. They are clear and legible from anywhere in the room.
   iii. You have the correct number of slides for the time allotted.
   iv. You know all the material well enough to answer any question on any aspect of anything you present.
   v. Divide your talk into three parts:
      - *The beginning:* an overview of the presentation and background or motivational material
      - *The main body:* what you have to say
      - *The conclusion:* a summary of what you have said, together with suggestions for future considerations

2. *Rehearse, rehearse, rehearse!* This is an absolute must. You must rehearse the actual presentation. You would not believe how different it is actually giving the talk you have prepared on paper. I always go through any presentation, in its entirety, in the allotted time, out loud to myself (or to anyone willing to listen), at least twice before I actually make the formal presentation to the intended audience (once the day before and once again a few hours before the presentation). Usually I do this in my hotel bedroom, in an empty conference room, or in my office—anywhere I can talk out loud without disturbing other people. The confidence boost you get from such a rehearsal is incredible—plus, you get to remove awkward phrases and adjust the material so that

it sounds better. Nothing will relax you more than knowing that you are well prepared and entirely confident about your presentation.

3. *Familiarize yourself with the room in which you will speak.* On the day of the presentation, arrive early and walk around the entire room (including where the audience will sit). Get familiar with the surroundings and your view during the presentation. Make sure all necessary equipment is in the room and in working order.

4. *Have a chat with someone in the room before the presentation begins.* Find someone in the room with whom you are acquainted, and have a friendly chat while the audience is seating itself. Not only will this relax you, but it will inform the audience that you are indeed relaxed, which adds to the appearance of a professional and polished presentation.

5. *Believe in yourself.* Get rid of any negative thoughts, and have nothing but 100-percent confidence in your abilities. Visualize the audience listening carefully to what you have to say and receiving your information with enthusiasm. Use the adrenalin rush as positive energy to project your voice and your personality throughout the room.

6. *Don't belittle yourself.* Don't say things that make you look unprofessional, for example, "You'll have to excuse me, this is only my second formal presentation" or "I don't know where this result came from. I just copied it out of a book. I'm not smart enough to do something like that," in an effort to make the audience feel better. It never works. All that happens is that the audience loses respect for your abilities and stops listening. Maintain 100-percent confidence and professionalism at all times. You are the expert; behave accordingly.

7. *The audience doesn't expect to be entertained.* In a technical presentation, the audience comes to learn something, not to be entertained. Don't try to tell jokes or try to make technical material funny. Concentrate on your message; that's what they want to hear above all else.

8. *Experience is unbeatable.* The more presentations you give, the more effective you will become as a speaker. Get involved in anything that will give you the opportunity to test yourself:
   - Take courses that allow you to make presentations.
   - Get constructive criticism from experts in public speaking. For example, ask your professor to criticize your performance and provide some tips to improve your effectiveness.
   - Lead your study group. Take the role of leader in your study group whenever you can.
   - Get involved in student government.
   - Join public speaking clubs.
   - Join debating clubs.
   - Read books and articles on giving formal presentations. (see, for example, *A Handbook of Public Speaking for Scientists and Engineers* [9] and *Tips on Talks* [10]).

As with writing and interpersonal skills, formal presentation skills take time to develop, so use your time at the university to make the most of every opportunity available to you to practice these skills.

9.2 DEVELOPING YOUR CREATIVITY

Creativity is basically defined as originality of thought or the ability to use the imagination to come up with new and innovative ideas. You can see why employers regard creativity as one of the most sought-after skills an engineer can possess. If an engineer is creative, not only is that engineer equipped (through his or her training) to turn ideas into reality, but also, the person is a source of new ideas, which often lead to new products or services, which in turn keep companies in business.

Can a person learn to be creative? Some people say no, claiming that creative tendencies are inherited rather than learned. Modern-day evidence, however, suggests that this may not indeed be the entire story—that creativity can, to a certain extent, be learned, developed, and continually improved.

The following are some general suggestions that you can implement today to develop your own creative-thinking skills:

- *Knowledge.* If you wish to be creative in engineering, you must know engineering. Do all that is necessary to be most effective in engineering study—to learn all that is offered you. This will equip you with the knowledge required to explore your ideas.

- *Maintain an interest in things outside engineering.* Keep your mind sharp by staying up-to-date and challenging or debating people on other issues outside your area of interest, for example, politics, your views on social and religious issues, government, and so on. Exposure to other ideas is an excellent way of nurturing your creative side.

- *Ask yourself "What if. . . . ?" Take a new look at old problems.* Let your mind wander. Whenever you solve a problem (from, for example, an assignment or the textbook), ask yourself what would happen if you changed the conditions of the problem.

- *Play with ideas.* If you have an idea, no matter how ridiculous or impossible it might seem, go with it. Play with it, twist it around, work it a little, and see what you come up with.

- *Brainstorm.* Get together with a group of colleagues, and discuss existing ideas or brainstorm new ones. Talking about your ideas will make them clearer and help them develop further.

- *Allow your subconscious to play its part.* The answer to a challenging problem is almost never obtained at first thought. Most people let the idea sit in their subconscious while they do or think about something else. The brain then goes on "autopilot" bringing the idea to the conscious mind at different times (sometimes even during sleep). Allow yourself to do this. Don't expect the solution immediately; let things simmer in your head for a while—let the idea develop. You must have had the experience of moments of inspiration occurring in the most unusual places. (For me, it's while I'm walking the dog or jogging—when my body is working and my mind is relaxing.) This is just your subconscious communicating with your conscious mind!

- *Take time to recharge.* It has been said many a time that the most creative people are dreamers, people who can lose themselves in their own imaginations. Take the time to relax, to dream, to ponder, to engage in things that slow the pace down a little. If you're rushing about the whole day, your mind is in a state of constant activity. Let it do what it likes for at least some time each day.

- *Ask questions.* In each of my classes, the students that ask the most questions are almost always the most creative students. They reach beyond what I tell them into the realms of the unsolved. You should never be afraid to ask questions.

> Curiosity is what drives creativity.

- *Cut out irrelevant details and get to the heart of the matter.* Creative people don't worry about *how* things will get done. (They believe this will come later, as they strive towards their goal), Instead they focus on the objective: what they would like to see happen, built, or developed.

The important thing to remember about creativity is that it is closer to a religious feeling than it is to science. It's more about belief, attitude, and approach than it is about facts and rules.

Problems

1. In a five-hundred word paper entitled "Why communication skills are an essential part of engineering," explain why communication skills play such a vital role in engineering.
2. What kind of documents do you think you will have to write as a practicing engineer? Make a list. From this list, identify which documents you can competently write now and which would require further development of your writing skills.
3. Write a letter to the dean of your faculty asking that he or she reconsider your grade in Scottish Culture 101, since you were sick during the final examination.
4. What are the strongest aspects of your writing skills? What are the weakest? Devise a strategy to deal with your weakest skills, and implement that strategy.
5. You have become so fed up with the substandard computing equipment in your engineering lab, that you decide to write a letter of complaint to the dean of the faculty. Write that letter. Make sure you describe your specific complaints and any suggestions you might have to overcome the problems.
6. Look at your last (final version) set of class notes. Edit them for clarity, grammar, and punctuation. Give them to someone to read over. Use any comments you receive to improve that particular set of notes further. Decide to apply these improvements to all subsequent notes you take in class.
7. Suppose that in two weeks time you will run out of money. Write a proposal for funding to anyone likely to give you money (for example, the Student Union, the office of financial aid, your parents, or local scholarship organizations) to complete your education. Make sure you explain why you need the money and why you should be given the money (i.e., what benefits the donor will get out of giving you the money), and provide a detailed account of how you propose to spend the money.
8. Write a letter to a local engineering firm asking for a summer job.
9. You need the use of a car for the evening. State how you would persuade one of your friends to lend you his or her car.
10. Write a summary of Chapter 9 of this book. Make your summary no more than two pages long. Let someone read it and tell you what they have learned from your summary. Does your summary capture the salient points of the chapter?
11. Have you ever given a formal presentation? If you have, write a one-page paper describing your experiences. If you haven't, write a one-page paper describing what you think it might be like to do so.

12. As student representative you are scheduled to attend the next academic planning committee meeting of your faculty. You will present the case for a consolidated (common) examination for all sections of the course Rigid Body Dynamics 250. Write your opening statement (i.e., why you think such an examination is necessary).

13. Plan a 20-minute seminar on a subject of your choosing. You should come up with the appropriate content for the allocated time, an opening statement, and a conclusion.

14. Give the seminar described in Problem 13 to a group of friends or to your study group.

15. List your strengths and weaknesses in each of the following areas:
    - Interpersonal communication
    - formal presentation

    Devise strategies for dealing with the weaknesses you have identified.

16. What are you going to do to ensure that when you graduate, you can perform effectively in meetings and conversation with professional clients?

17. How would you prepare for a job interview? Practice these techniques by having a friend interview you for some fictitious summer job.

18. Have you ever done something that you regard as being *creative?* Describe the experience.

19. What do you think makes for creativity in an engineer? What would you do to teach creative skills to engineering students?

20. Why is creativity so important to employers of engineering graduates? Write a one-page paper that answers this question in detail.

21. Do you think you are creative? Why or why not?

22. Devise a strategy for improving your creativity. Implement the strategy.

23. How would you advise your professors to teach creativity? What do you think will work? Make some suggestions, and ask a particular professor if he or she would implement them in class.

10

Looking to the Future: What's after Graduation?

Our discussion so far has been concerned with maximizing performance in engineering study. The ultimate goal has been the successful completion of your undergraduate degree in engineering, the foundation of your engineering education. Upon completing your degree, you will have a variety of options for what to do next. Basically, the choice comes down to one of the following two possibilities:

- Go to work as a practicing engineer.
- Continue your study towards a graduate degree.

In this chapter, we will take a brief look at both options.

10.1 GOING TO WORK AS A PRACTICING ENGINEER

In Chapter 2, we discussed various careers available to engineers specializing in different disciplines. Employment in these careers can be found in the private and public sectors of the economy, including overseas employment:

- *Private sector.* Employment in the private sector includes:
  —*Consulting engineering*
  —*Project management*
  —*Related areas.* Many firms recruit engineering graduates to work in nonengineering areas

OBJECTIVES

In this chapter you will learn:

- About the opportunities available to you after you graduate in engineering.
- How to go about finding a job.
- About working as a practicing engineer.
- About different types of graduate degrees in engineering.
- How to continue your studies towards a graduate degree in engineering.

because of the valuable skills they have acquired as university students. For example, management consulting firms, accounting firms, and insurance and financial companies all value the problem-solving, analytical, and creative aspects of an engineering education.

—*Teaching at the secondary level.* Education is another area that often interests engineering graduates, for example, teaching at the secondary level. In this case, further study in the form of at least a master's degree would be necessary

- *Public sector.* Graduate engineers can work for any level of government in government-owned industries and organizations, for example, water, gas, electricity, transportation, and so on.

- *Working overseas.* Many large multinational engineering (and nonengineering) firms, as well as numerous government agencies, have offices all over the world. Consequently, they offer opportunities for graduates to travel and work in many different countries.

Some of my more recent graduates are employed as follows:

Louisa   A 1992 graduate in mechanical engineering, Louisa works in the private sector as a project manager with a large oil company.

Calvin   Calvin, who graduated in 1990 with a degree in electrical engineering, now works in Hong Kong as a computer engineer for a large financial corporation.

Sandra   Sandra graduated in 1994 in mechanical engineering and now works for a biomedical company in Canada.

Robert   A 1991 graduate in civil engineering, Robert started his own computer company, of which he is now president and CEO.

The important thing to remember is that, even though you choose to enter the work world rather than pursue a graduate degree, your bachelor-of-science degree in engineering is not the end of your engineering education. On the contrary, it's just the beginning. Your time at the university has not only allowed you to become a skilled, independent thinker and learner with sought-after analytical and problem-solving abilities, but has also given you a road map for lifelong learning. In other words, you have learned how to learn! Consequently, as a practicing engineer, you will be involved in continuing education as you:

- Maintain or renew your career.
- Continually upgrade your skills.
- Stay in touch with advances in technologies.

There is no doubt that the next millennium will see a vastly changed environment for the engineer. To prepare for this, you must realize that continuing education is the key. You cannot afford to stand still. You must continually improve yourself. That's what an engineering education allows you to do. It is less about learning a specific set of skills to do a specific job than it is about training your mind and giving you the ability to learn and adapt to any situation. Your first job will be only the first step in this process.

To end this section, I'd like to give a few useful tips about finding a job. The following list presents what I have learned over the years in this respect:

1. *I've yet to meet someone who has found a job by replying to a newspaper advertisement.* In my experience, most jobs arise through word of mouth from *personal contacts.* Get to know people in the industry of your choice through friends, by attending parties, through your professors, during your time at the university (a good reason for attending meetings, conferences, etc.), or through summer or temporary employment. Or just approach companies directly.

   For example, the following list shows the various companies I have been with or positions I have held and how they came about:

   i. Smith's Industries Aerospace and Defence Systems Company: personal contact

   ii. Marconi Instruments: wrote to the company directly

   iii. Ferranti Defence Systems: wrote to the company directly

   iv. Postdoctoral position at the University of Alberta: personal contact

   v. Director of Mathematics Resource Center: personal contact

   vi. Positions in mathematics departments in various countries: contacts made at conferences and meetings.

   vii. Present faculty position in engineering (personal contact made through a friend)

2. *Apply hard and apply fast.* When you have decided to apply for jobs, apply for lots of them at the same time. This will increase the likelihood of receiving interviews or job offers and prevent the job-hunting process from dragging on.

3. *Prepare an excellent resume or curriculum vitae (CV).* Exaggerate, by all means, but never lie. Employers will check references. More and more people are being caught as competition for jobs increases. Keep the CV brief, presenting only the main points. The interview is the place to expand on detail. Believe me, many superbly qualified people are overlooked because those doing the hiring cannot be bothered reading through a 50-page CV.

4. *An interview is an open door.* Getting an interview from the efforts just described is an invitation to come and impress the relevant people. *Prepare!* You should approach an interview as you would a formal presentation. (See Chapter 9.) Learn as much as possible about the company (and the people conducting the interview), and present yourself (and your information) in the best possible light. Also, *ask questions!* Interviewers love to be interviewed. (This has certainly been my experience on both sides of the fence.)

5. *Stay on good terms with important and influential people.* You never know when you might need to use them as references.

6. *Qualifications are great, but they are one dimensional.* What matters is the person holding the qualifications. In any interview I have conducted, what always impresses me most is personality, attitude, and drive—possessed by people who are hungry for success. These people often demonstrate the potential to learn anything. I'd rather hire a hungry smart person with no qualifications than someone with advanced degrees, but with no drive or ambition. Use the interview to show people who you are. Don't hold back!

7. *Most employers want someone who can demonstrate the ability to learn and think independently.* They like to hire people who can be left alone to get on with the job. Sometimes this may seem unfair (particularly if you are expected

to perform in an area that is unfamiliar to you and you need a period of training), but it's reality, so don't give the impression that you will need lots of help and training.

8. *You can do anything!* If you really want a particular job, do what you have to do to get it, and worry about how you will do things later. For example, I once accepted a job teaching structured computing languages to business students. I had never studied computing, but I really needed the job, and I knew I was able and had time to learn the subject matter.

Finally, in Chapter 9, we discussed the importance of good writing skills. This is paramount in applying for jobs. Your prospective employer knows you only from what you give or tell him. If you write a letter riddled with errors in grammar, punctuation, and spelling, what chance do you have of going further? If you are in doubt, get someone you can trust to check what you have written.

10.2 CONTINUING YOUR STUDIES TOWARDS A GRADUATE DEGREE

There are basically three degree options available to you if you choose to enter graduate school.

Master-of-Science Degree (M.S.) in Engineering   The master-of-science degree is offered in various formats at different institutions. Normally, it takes one to two years to achieve and covers differing amounts of course work, project work, and thesis-based research. For example, my own institution offers two options. One option involves taking five graduate courses plus presenting an acceptable thesis based on the student's research (an oral defense of the thesis is mandatory), while the other option involves taking eight graduate courses plus a completing project of relatively short duration. You can check which options are available at your particular institution simply by consulting the university calendar or by contacting a graduate advisor. Students receiving the master-of-science degree may choose to continue studies toward a doctoral degree or to enter engineering practice.

Doctoral Degree (Ph.D.) in Engineering   This is a research degree and the highest degree available in engineering. It normally takes anywhere from three to five years to earn. The main component is a thesis based on the student's own research (under the guidance of an appointed supervisor), but there is also a component of course work. The choice to enter a doctoral program must be a personal one, based on your own specific preferences and desires to conduct research in an area of your choosing. Perhaps the most important qualities required to complete this degree are curiosity, perseverance, motivation, and a genuine interest in pursuing research in the area that interests you most. Most people entering a Ph.D. program do not do so with the expectation that they will improve their employment prospects (unless they intend to pursue a career in academia, in which case a Ph.D. is certainly necessary). Some students enter a Ph.D. program after completing an M.S. degree that includes a component of research. In this way, they get a chance to sample the lifestyle and gain more information about the type of work expected in the Ph.D. program.

Degrees in Related Disciplines   Many engineering graduates choose to study advanced degrees not specifically in engineering, but in disciplines related to their chosen careers.

For example, one of the more common graduate degrees taken by engineers wishing to pursue more managerial-type careers is the master of business administration (M.B.A.). There are also master's degrees in, for example, mathematical sciences, mathematical finance, meteorology, and bioengineering, and many more degrees of an interdisciplinary nature. Taking a degree of this type usually opens doors to other degrees at an even higher level. For example, one of my former colleagues taught mathematics for many years, yet began his career with a B.S. degree in electrical engineering. Engineering graduates are also welcomed into law school and medical school, mainly because of the logical thinking and problem-solving skills developed over the course of earning an undergraduate engineering degree.

Many engineering graduates choose to take on further study after having had a few years work experience. This is exactly what I did. I found that my time in industry gave me:

- A better idea of what I wanted to do.
- A better idea of the different career paths and options open to me.
- A better appreciation of the area in which I was interested in specializing.
- An opportunity to put into practice some of the things I had learned in university.
- A chance to see if the grass was indeed greener on the other side.
- A chance to make some money.

Although these are all good reasons for working before returning to graduate school, I found that the biggest drawback was the time taken away from what I like to call the undergraduate mentality. After having savored a lifestyle in which you have money in your pocket, no examinations, and the freedom to do the things you (more or less) want to do, it's difficult to go back to being a full-time student, which brings me to the question of money.

MONEY MATTERS

Graduate study is much more like having a job than studying at the undergraduate level. Usually, at the graduate level, you are paid (often monthly) to pursue research. For example, you can earn money from a teaching assistantship or from a research assistantship, and you can apply for fellowships, student awards, or scholarships and enter various competitions for funding. My own graduate students obtain money from these and other sources. (For example, I apply on their behalf for money from a federal granting agency.) This stipend is expected to pay for everything, including tuition, fees, and living expenses. However, the payments are nowhere near what you would earn in industry. They are designed specifically to provide enough for you to devote all of your time to your studies so that you can graduate and enter the work world in the shortest possible time.

Problems

1. Write a 500-word paper entitled *"Work versus Graduate Study: Which One Is for Me?"* and state the advantages and disadvantages of each.

2. Write down your top five choices for a career. For each career possibility, list five companies that might employ you.

3. What are the most important skills you think are required to succeed in the work world. Which do you think will develop as a result of your engineering education?

4. What are the most important skills you think are required to succeed in graduate studies. Which do you think will develop as a result of your engineering education?

5. Why do you think people change universities when they decide to pursue graduate study? Why don't they just remain in familiar surroundings? Write a one-page paper answering these questions.

6. List 10 qualities that you think are important for successful research towards a Ph.D. degree. How many do you have? What would you do to develop those that you don't have, yet deem necessary.

7. Investigate the graduate programs available in engineering at your university. Are there any *new* programs designed specifically to meet the demands of modern-day industry?

8. How do you view your first job after graduation? Is it the job you intend to pursue for the rest of your life? Must it be exactly want you want to be doing in 10 years, or is it just a stepping-stone to the unknown. Write a one-page paper on "My First Job after Graduation: What I Expect."

9. Suppose, after graduation, you were offered a job that paid lots of money, was in a great location, and came with a company car and excellent benefits, but you hated the day-to-day requirements of the job. What would you do? Write a one- page paper entitled "The Importance of Enjoying My Work," and in it answer these questions.

10. List 10 reasons you might pursue graduate studies immediately after completing your undergraduate degree in engineering.

11. List 10 reasons you might pursue graduate studies two years after completing your undergraduate degree in engineering (and having been employed by two different companies, each for a period of one year).

References

[1] R. M. Felder & J. E. Stice, *National Effective Teaching Institute Manual,* ASEE, Anaheim, CA, 1995

[2] R. M. Felder, *An Engineering Student's Survival Guide,* illustrated by Prentis Rollins, reprinted from Chapter 1, fall 1993, pp. 42–44, © 1993, American Institute of Chemical Engineers (AICHE), reprinted with permission. New York, N.Y.

[3] D. W. Johnson, R. T. Johnson, and K. A. Smith, *Cooperative Learning,* ASHE-ERIC Higher Education Report No. 4, George Washington University, Washington, DC, 1991.

[4] P. Schiavone, *How to Succeed in Calculus, from Essential Prerequisites to Practice Examinations,* Prentice-Hall Canada, Scarborough, Ontario, 1997.

[5] P. Schiavone, *Introductory Ordinary Differential Equations,* Prentice-Hall Canada, Scarborough, Ontario, 1998.

[6] A. Robbins, *Unlimited Power,* Random House, New York, 1986.

[7] A. Robbins, *Awaken the Giant Within,* Simon & Schuster, New York, 1991.

[8] S. R. Covey, *The Seven Habits of Highly Effective People,* Simon & Schuster, New York, 1989.

[9] P. Kenny, *A Handbook of Public Speaking for Scientists and Engineers,* A. Hilger Ltd., Bristol, U.K., 1983.

[10] R. M. Felder, *Tips on Talks,* Department of Chemical Engineering, North Carolina State University, Raleigh, North Carolina 27695-7905. See also <http://www2.ncsu.edu/unity/lockers/users/f/felder/public/Papers/speakingtips.htm>.

Index